Fundamentals of
Process Control Theory
Third Edition

Fundamentals of Process Control Theory
Third Edition

by Paul W. Murrill, Ph.D.

Notice

The information presented in this publication is for the general education of the reader. Because neither the author nor the publisher have any control over the use of the information by the reader, both the author and the publisher disclaim any and all liability of any kind arising out of such use. The reader is expected to exercise sound professional judgment in using any of the information presented in a particular application.

Additionally, neither the author nor the publisher have investigated or considered the affect of any patents on the ability of the reader to use any of the information in a particular application. The reader is responsible for reviewing any possible patents that may affect any particular use of the information presented.

Any references to commercial products in the work are cited as examples only. Neither the author nor the publisher endorse any referenced commercial product. Any trademarks or tradenames referenced belong to the respective owner of the mark or name. Neither the author nor the publisher make any representation regarding the availability of any referenced commercial product at any time. The manufacturer's instructions on use of any commercial product must be followed at all times, even if in conflict with the information in this publication.

10 9 8
Printed December 2011

ISA
67 Alexander Drive
P.O. Box 12277
Research Triangle Park
North Carolina 27709

Library of Congress Cataloging-in-Publication Data

Murrill, Paul W.
    Fundamentals of process control theory / by Paul W. Murrill. -- 3rd ed.
       p.   cm.
    ISBN 1-55617-683-X
    1. Process control.   I. Title.
  TS156.8.M87 2000
  670.42'75—dc21   99-27991
                                                           CIP

To Nancy

TABLE OF CONTENTS

Preface to the Third Edition

Fundamentals of Process Control Theory was written in 1981 as a prototype for ISA's Independent Learning Module publication series, and it rapidly became the best-selling book ever published by ISA. With the publication of the second edition in 1991, the strong popular acceptance continued, and the book has approached "classic" status. This is especially satisfying for an author.

This textbook is designed for independent self-study. It is for the practicing engineer, first-line supervisor, or senior technician. College, university, and technical school students will also find the material appropriate. *Fundamentals of Process Control Theory* is designed to teach the basic principles of process automation and demonstrate how these principles are applied in modern industrial practice. Some knowledge of mathematics is necessary, of course, but I have made efforts to prevent the mathematics from being a barrier to study by those without strong math skills. The material is designed as an introductory or first-level course. A quick review of the table of contents will provide an insight into the specific topics covered.ISA

This book is intended to be both theoretical and practical—that is, to show the basic concepts of process control theory and how these concepts are used in daily practice. This is a book about fundamentals, concepts, ideas, principles, theory, and behavior. It is not about hardware and software. To some extent, however, we all know that hardware and software will dictate what "theory" can be used and useful. Thus, the actual implementation of process control theory changes as hardware and software change, and this march of progress has made a third edition necessary. I hope this new edition proves useful to you, the student.

An effort has been made to make this presentation consistent with the various standards and practices used throughout the various worlds of process control and instrumentation. I have attempted to ensure that no significant inconsistency exists with the standards relating to terminology, especially *Process Instrumentation Terminology*, ANSI/ISA-S51.1-1979 (R 1993), and that is reflected in both the text and the Glossary given in Appendix B. Some minor inconsistencies may be noted with ANSI/ISA-S5.1-1984 (Reaffirmed 1992) *Instrumentation Symbols and Identification* (referenced and excerpted in Appendix A), in some of the figures of the presentation, but in no case have I found this to lead to confusion for students.

My special thanks goes to all who made the first and second editions of this book so successful. It is my hope that the revisions for the third edition, many of which were suggested by you, will make it even better.

Paul W. Murrill, July 1999

Unit 1:
Introduction
and Overview

UNIT 1

Introduction and Overview

Welcome to *Fundamentals of Process Control Theory*. The first unit of this self-study program provides you with the information you will need to proceed through the course.

Learning Objectives — When you have completed this unit, you should:

A. understand the general organization of the course

B. know the course's objectives

C. know how to proceed through the course

1-1. Course Coverage

This is an introductory book on the fundamental principles of automatic process control. This course covers

a. the basic theoretical concepts of automatic process control.

b. how these basic theoretical concepts are applied in modern industrial practice.

The scientific principles that process control is based upon are unchanging. There is, however, considerable variation in the hardware and software available from vendors and in the application and practice of these control principles from one industry to another. The material presented in this course is generally oriented toward the modern practitioner in such processing industries as petroleum, petrochemical, chemical, pulp and paper, mining, power, drug manufacturing, and food processing.

There is very little in this book about specific hardware and software; instead, the focus is on the fundamentals of theory, concepts, behavior, relationships, and ideas. This course will focus on process control and will not provide an extensive discussion of individual measurement techniques (such as flow measurement or temperature measurement) except to illustrate the techniques and to show how measurement affects control.

No attempt is made in this book to provide exhaustive analysis on how to control a specific unit operation or process (such as, for example, heat exchangers or distillation columns). Instead, the focus is on the application of the general principles of control theory.

1-2. Purpose

The purpose of this book is to present the basic theoretical principles of automatic process control in easily understandable terms and to illustrate and teach you how these principles are used in modern industrial applications. This is neither solely a theoretical course nor solely a practical course--it is both! The book's purpose is to show the theoretical concepts and principles in day-to-day commercial and industrial situations and, in doing so, to show that this theory is quite practical.

1-3. Audience and Prerequisites

This book is designed for those who want to work independently and who want to gain a basic introductory understanding of automatic process control. The material will be useful to engineers, first-line supervisors, and senior technicians who are concerned with the application of process control. The course will also be helpful to students in technical schools, colleges, or universities who wish to gain some insight into the practical concepts of automatic process control.

There are no elaborate prerequisites for this course, though an appreciation for industrial concerns and their philosophies will be helpful. In addition, it is inevitable that particular parts of the presentation will involve some mathematics. However, the student does not need to be intimately familiar with such mathematics to appreciate the control concepts that are presented and applied here. Quite often, mathematics becomes one of the barriers that prevent people from understanding and actually using process control theory; in this textbook I have attempted to minimize such barriers.

1-4. Study Material

This textbook is the only study material required. It is an independent, stand-alone textbook that is uniquely and specifically designed for self-study.

1-5. Organization and Sequence

This book is divided into sixteen separate units. The next five units (Units 2-7) are designed to teach the student basic feedback control concepts and the functional components that are used in modern industrial applications. Following these five units, there are two units (Units 8 and 9) that introduce the student to the dynamic behavior and tuning of process control systems. The next five units (Units 10-14) give the student an introduction to more advanced control techniques and concepts. The last two units (Units 15-16) present control system architecture and new directions for process control.

Because the method of instruction used in this book is self-study, you select the pace that is best for you. Each unit is designed to have a consistent format in which a set of specific learning objectives is stated at the very beginning of the unit. Note these learning objectives carefully; the material that follows is keyed to these objectives. Some units contain numbered example problems to illustrate specific concepts, and at the end of most units you will find exercises to test your understanding of the material. Where exercises are not given, it is because the material of the unit is not quantitative in nature. The solutions to all exercises are given in Appendix C, and you should be sure to check that your solutions are correct.

This book belongs to you; it is yours to keep. You are therefore encouraged to make notes in the book and to take advantage of the ample white space provided on every page for this specific purpose.

1-6. Course Objectives

When you have completed this entire book, you should:

a.   understand the basic theoretical concepts of feedback control

b.   understand the functional role of the specific hardware components used in process control applications

c.   have an appreciation of process dynamics and the tuning of industrial control systems

d.   have an appreciation of advanced control techniques such as cascade control, ratio control, dead time control, feedforward control, and multivariable control

e.   understand how digital processing capabilities are used in process control applications

f.   have some appreciation of overall process control philosophies and strategies

As we mentioned earlier, in addition to these overall course objectives a specific set of learning objectives for each unit is given at the beginning of each unit. These objectives are intended to help direct your study of that individual unit.

1-7.   Course Length

One basic premise of self-study is that students learn best if they proceed at their own comfortable pace. As a result, there will be a significant variation in the time individual students take to complete this book. Most students will complete this course in fifteen to eighteen hours, though previous experience with the material and personal capabilities will do much to vary this time.

You are now ready to begin your detailed study of the basic concepts of automatic process control theory. Please proceed to Unit 2.

Unit 2:
Basic Control Concepts

UNIT 2

Basic Control Concepts

This unit introduces the basic concepts encountered in automatic process control. Some of the basic terminology is also presented.

Learning Objectives — When you have completed this unit, you should:

    A.   be able to explain the meaning of the following terms:

          1.   controlled quantities
          2.   disturbances
          3.   manipulated quantities

    B.   understand the basic concept of feedback control

    C.   understand the basic concept of feedforward control

    D.   have a general overview of process automation

2-1. Control History

The first well-defined use of feedback control seems to have been James Watt's application of the flyball governor to the steam engine in about 1775. As a matter of interest, most of the early applications and theoretical investigations of feedback control were associated with governors, and these usually were in industrial applications. Broader use of automatic control began to be made in the late 1920s, and the first general theoretical treatment of automatic control was published in 1932. The growth in industrial usage has been steady and strong.

Many new technologies have been applied to process control hardware as the industrial use of automation techniques has developed and matured in the past seventy years. An important example of this was the application of digital computer and microprocessor capabilities to process control in the 1960s. As a result, process automation received a significant and very special boost in technology. Today, many industries allocate in excess of 10 percent of their plant investment capital outlays for instrumentation and control. This percentage has doubled over the past thirty years and shows no signs of diminishing.

The underlying theory of automatic control has also developed rapidly, and a firm and broad foundation of understanding has been created. Today's applications are based on this foundation. Many modern practitioners encounter difficulty, however, applying well-defined mathematical theories of automatic process control. This difficulty is quite natural, but much of the problem is due to the fact that teachers do not focus sufficient attention on illustrating theoretical principles by studying day-to-day industrial applications. The principle purpose of this book is to alleviate this problem by showing the actual use of control theory in practice.

2-2.   The Variables Involved

To understand automatic process control, you must first fix in your mind three important terms that are associated with any process: *controlled quantities*, *manipulated quantities*, and *disturbances*. These are illustrated in Fig. 2-1. The controlled quantities (or *controlled variables*) are those streams or conditions that the practitioner wishes to control or to maintain at some desired level. These may be flow rates, levels, pressures, temperatures, compositions, or other such process variables. For each of these controlled variables, the practitioner also establishes some *desired value*, also known as the *set point* or *reference input*.

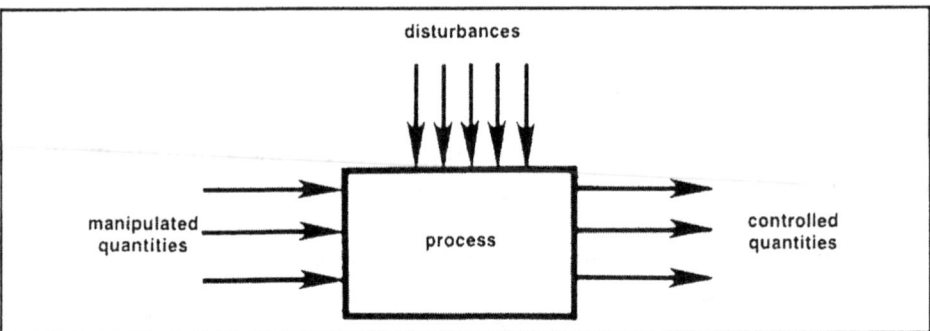

Figure 2-1. The Variables Involved

For each controlled quantity, there is an associated manipulated quantity or *manipulated variable*. In process control this is usually a flowing stream, and in such cases the flow rate of the stream is often manipulated through the use of a control valve. Disturbances enter the process and tend to drive the controlled quantities or controlled variables away from their desired, reference, or set point conditions. The automatic control system must therefore adjust the manipulated quantities so that the set point value of the controlled quantity is maintained in spite of the effects of the disturbances. Also, the set point may be changed, in which case the manipulated variables will need to be changed to adjust the controlled quantity to its new desired value.

Fig. 2-2 shows a typical home heating system. In such a system, the controlled variable is room temperature. (Your intent, of course, is to maintain creature comfort in the room, and you control "comfort" by controlling a variable that can be measured easily, such as temperature.) A number of disturbances cause room temperature to vary, for example, outside ambient temperature, the number of people in the room, the type of activity taking place in the room. The automatic control system is designed to manipulate the fuel flow to the furnace in order to maintain room temperature at its desired value or set point in spite of the various disturbances.

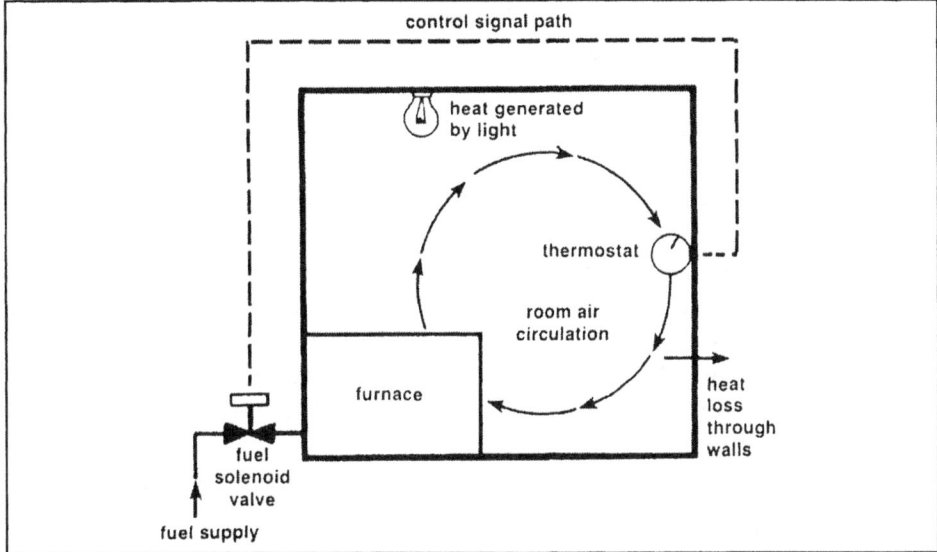

Figure 2-2. A Home Heating System

2-3.   Typical Manual Control

Before studying automatic process control, it is helpful to spend a moment or two reviewing a typical manual operation. This is illustrated in Fig. 2-3, which shows a process with one controlled quantity. On the stream leaving the process, there is an indicator to provide the operator with information on the current actual value of the controlled variable. The operator is able to inspect this indicator visually and, as a result, to manipulate a flow into the process to achieve some desired value or set point of the controlled variable. The set point is, of course, in the operator's mind, and the operator makes all of the control decisions. The problems inherent in such a simple manual operation are obvious.

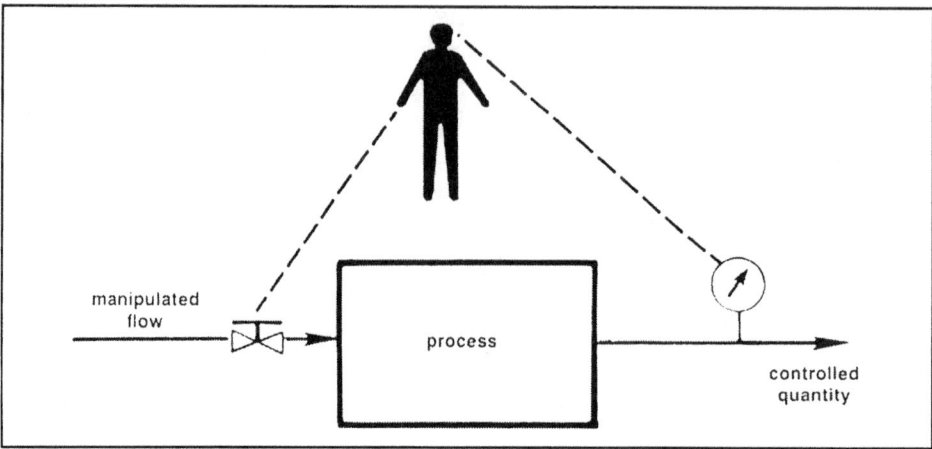

Figure 2-3. Typical Manual Control

2-4.   Feedback Control

The simplest way to automate the control of a process is through
conventional feedback control. This widely used concept is illustrated in
Fig. 2-4. Sensors or measuring devices are installed to measure the actual
values of the controlled variables. These actual values are then transmitted
to feedback control hardware, and this hardware makes an automatic
comparison between the set points (or desired values) of the controlled
variables and the measured (or actual) values of these same variables.
Based on the differences ("errors") between the actual values and the
desired values of the controlled variables, the feedback control hardware

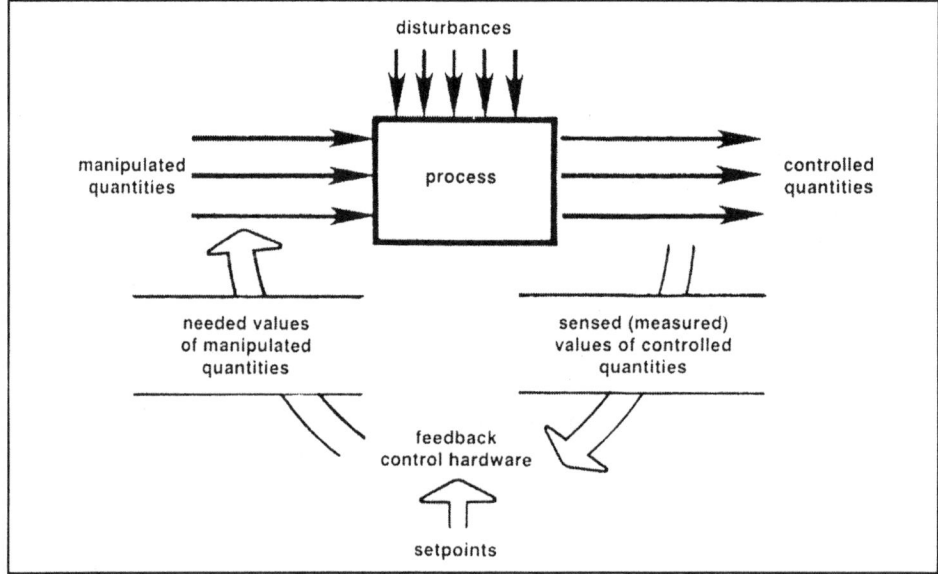

Figure 2-4. Feedback Control Concept

calculates signals that reflect the needed values of the manipulated variables. These are then transmitted automatically to adjusting devices (typically control valves) that manipulate inputs to the process.

The beauty of feedback control is that the designer does not need to know in advance exactly what disturbances will affect the process, and, in addition, the designer does not need to know the specific quantitative relationships between these disturbances or their ultimate effects on the controlled variables. The control hardware is used in a standard format, and all feedback control loops tend to reflect the general conceptual framework illustrated in Fig. 2-4. To a very significant extent, this standard pattern exists regardless of the specific nature of the process or the controlled variable involved.

The particular hardware used in a loop and the particular matching of one hardware piece to another is an important responsibility for the designer, but the overall control strategy is always the same in feedback control. Such feedback control is the simplest automatic process control technique that can be used, and it represents the basis for the vast majority of industrial applications.

2-5.   Manual Feedforward Control

Feedforward control is much different in conception from feedback control. A manual implementation of feedforward control is illustrated in Fig. 2-5. As a disturbance enters the process the operator observes an indication of the nature of the disturbance, and based on that entering disturbance the operator adjusts the manipulated variable so as to prevent any ultimate change or variation in the controlled variable caused by the disturbance. The conceptual improvement offered by feedforward control is apparent. Feedback control worked to eliminate errors, but feedforward control operates to prevent errors from occurring in the first place. The appeal of feedforward control is obvious.

Feedforward control does escalate tremendously the requirements of the practitioner, however. The practitioner must know in advance what disturbances will be entering the process, and he or she must make adequate provision to measure these disturbances. In addition, the control room operator must know specifically when and how to adjust the manipulated variable to compensate exactly for the effects of the disturbances. If the practitioner has these specific abilities and if they are perfectly available, then the controlled variable will never vary from its desired value or set point. If the operator makes some mistake or does not anticipate all of the disturbances that might affect the process, then the controlled variable will deviate from its desired value, and, in pure feedforward control, an uncorrected error will exist.

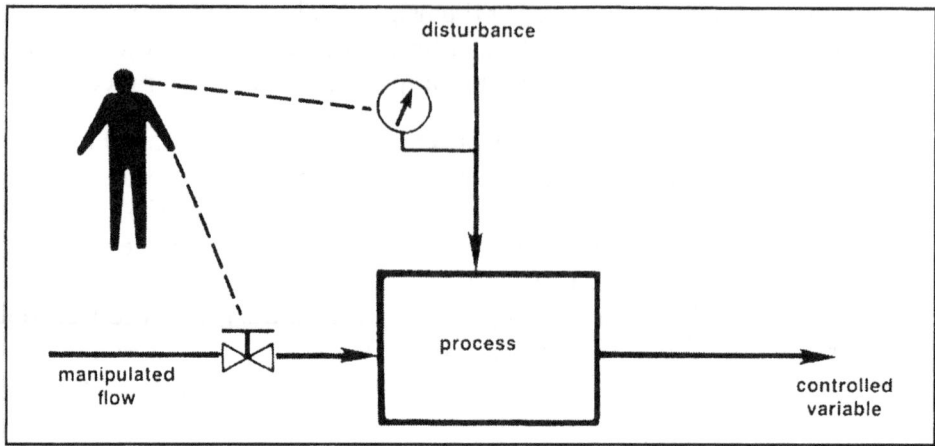

Figure 2-5. Manual Feedforward Control

2-6.   Automatic Feedforward Control

Fig. 2-6 shows the general conceptual framework of automatic feedforward control. Disturbances are shown entering the process, and sensors are available to measure these disturbances. Based on these sensed or measured values of the disturbances, the feedforward controllers then calculate the needed values of the manipulated variables. Set points that represent the desired values of the controlled variables are provided to the feedforward controllers.

It is clear that the feedforward controllers must make very sophisticated calculations. These calculations must reflect an awareness and understanding of the exact effects that the disturbances will have on the controlled variables. With such an understanding, the feedforward

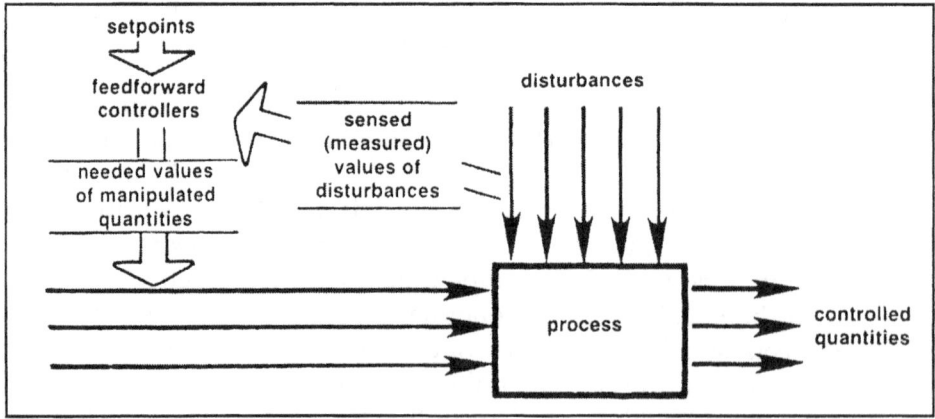

Figure 2-6. Feedforward Control Concept

controllers are able then to calculate the exact amount of manipulated quantities required to compensate for the disturbances. These computations also imply a specific understanding of the exact effects that the manipulated variables will have on the controlled variables. If all of these mathematical relationships are readily available, then the feedforward controllers can automatically compute the variation in manipulated flows that is needed to compensate for variation in disturbances. The escalation in the theoretical understanding required is obvious. Feedforward control, while conceptually more appealing, significantly escalates the technical and engineering requirements of the designer and practitioner. As a result, feedforward control is usually reserved for only a very few of the most important loops within a plant. While the number of applications is small, their importance is quite significant.

Pure feedforward control is rarely encountered, and it is more common for a process to have combined feedforward and feedback control loops. This will be illustrated in Unit 11 where feedforward control is discussed in greater detail.

2-7. Process Control *and* Process Management

Process automation is commonly used to derive the maximum profitability from a process. In the previous sections of this unit there was an implicit assumption that we knew the desired values (typically "desired" in order to achieve maximum profitability) for the controlled quantities. Once these desired values are known, automation techniques are applied to achieve and/or maintain these desired values or set points. Upon reflection, however, it can be seen that some of the most significant questions associated with the profitability of a process are those that must be answered to determine the desired values. This is basically the supervisory or management function, and quite often it is left for the human operator to determine. But, in recent years, with the significant advances in process automation many of these supervisory or management functions have themselves become automated, and the ability to achieve technological solutions and hardware answers for such management questions is a significant part of the modern control scene.

In a particular process, as the level of automation is increased, most of the initial steps involve using conventional process control (such as feedback control). However, as the level of automation increases more and more of the automation is associated with process management. This is illustrated in Fig. 2-7.

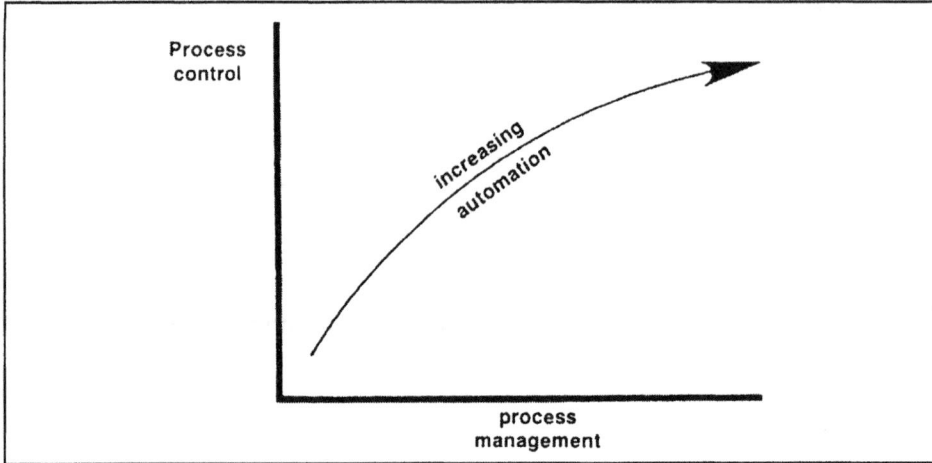

Figure 2-7. Process Control *and* Management

The combination of these two phenomena—process control and process
management—must be reflected in our overall understanding and
appreciation of process automation. A more detailed comparison of these
two subjects will be presented in Section 16-1.

EXERCISES

2-1. *Consider an electric oven in a typical modern kitchen. Identify the
controlled variable, the manipulated variable, and the disturbances.*

2-2. *Consider an automatic gas-fired, home hot-water tank. Identify the
controlled variable, the manipulated variable, and the disturbances.*

2-3. *Imagine you own a backyard swimming pool! Describe a manual control
system to measure pH and to add an acidic solution to adjust pH. Define
the controlled variable, the manipulated variable, and the disturbances.*

2-4. *Now automate the control of your swimming pool! Assume you have a tank
of acid solution to pump into your pool to control pH; use feedback control.*

2-5. *The "Cruise Control" feature used to control speed in an automobile is a
good example of feedback control. Outline its operation in terms of feedback
control.*

2-6. *Consider a gas-fired, home hot-water tank being used in a house that uses a
lot of hot water. This heavy usage, of course, is the disturbance or load on
the tank. Using a diagram, show how such a tank could be controlled using
feedforward control.*

2-7.   *For the hot-water tank in Exercise 2-6, could you use combined feedback and feedforward control? Using a diagram, show how.*

2-8.   *Consider a "moon shot" by NASA to send a rocket payload to the moon as a preprogrammed feedforward control problem. Analyze such a shot in feedforward control terms. What is the significance of the midcourse correction?*

2-9.   *In many cases today, a home heating system's thermostat is coupled to a microprocessor so the temperature set point may be managed so as to save energy. Analyze such a system in the terms of process control and process management.*

Unit 3:
Functional Structure of Feedback Control

UNIT 3

Functional Structure of Feedback Control

The general concept of feedback control was presented in Unit 2. Now we will take this general concept and reduce it to a functional layout for a single feedback control loop.

Learning Objectives — When you have completed this unit, you should:

A. understand the functional layout for a single feedback control loop

B. be able to explain the components of block diagrams

C. appreciate the mathematical structure of a single feedback control loop

3-1. A Single Feedback Control Loop

Any given process will have a number of different controlled variables, and for each controlled variable you must select an associated manipulated variable. So far we have only discussed this in the broadest and most general terms. But in practice a specific controlled variable is paired to a specific manipulated variable through appropriate feedback control hardware. This is done in the manner illustrated in Fig. 3-1.

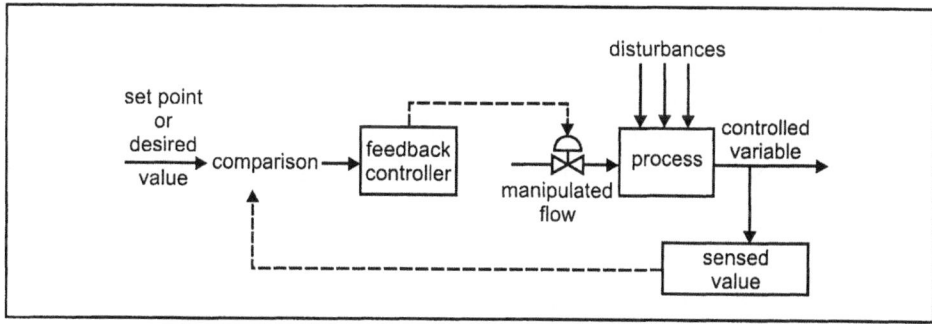

Figure 3-1. A Single Feedback Loop

The controlled variable is "sensed" or measured through appropriate "sensor" instrumentation, and this measured value of the controlled variable is then compared to the desired value of the controlled variable (the set point). The *difference* between these two (the error) is used as input for the feedback controller. Some would define this as the *deviation*, which is defined as the variation from the desired value.

Based on the error or deviation, the controller then calculates a signal to adjust the manipulated variable. Since the manipulated variable is normally a flow, the output of the feedback controller is usually a signal to a control valve, as illustrated in Fig. 3-1. While all of this is happening in a continuous fashion, disturbances may enter the process and tend to drive the controlled variable in one direction or another. The single manipulated variable is used to compensate for all such changes produced by the various disturbances. In addition, if there are changes in the set point the manipulated variable is also changed accordingly to produce the needed change in the controlled variable.

In a functional sense, all feedback control operates as illustrated in Fig. 3-1. Study this loop layout very, very carefully.

3-2.   Block Diagrams

To have a consistent way of representing control systems pictorially, it is useful to take advantage of block diagrams. Block diagrams are a simple, symbolic, graphical tool commonly used in automatic control. Block diagrams have two basic symbols; the first is a circle:

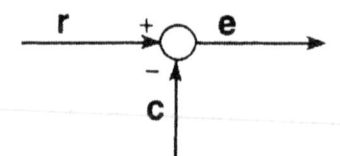

The arrows entering and leaving the circle are not vectors. They do, however, represent variables and actually represent the flow of information. The head of each arrow has an algebraic sign associated with it, either plus or minus. If no sign is present, an implied plus is intended. The small circle is really a simple way to represent algebraic addition or subtraction. The symbol shown in the diagram just presented represents the algebraic equation $r - c = e$.

The other symbol of a block diagram is, in fact, a block with one arrow entering and one arrow leaving:

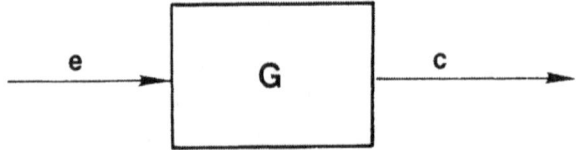

This is the way the algebraic operations of multiplication and division are symbolically presented. The output of the block is simply equal to whatever is contained within the block multiplied by the input. The block just shown represents the equation $c = Ge$.

Block diagram symbols may be combined into networks. Fig. 3-2 shows a block diagram for a very simple negative feedback control loop; it is "negative" because of the subtraction of the signal being fed back. This figure actually represents a combination of the two symbols shown earlier.

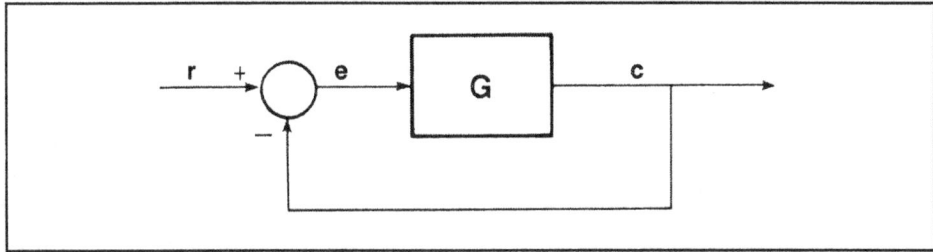

Figure 3-2. Block Diagram of a Single Loop

Block diagrams are used consistently throughout this book to pictorially present automatic control principles and applications.

3-3.   The Functional Layout of a Feedback Loop

The general layout and structure of the single feedback control loop now needs to be expanded to more closely represent the way feedback control is used in practice. Such a functional layout is illustrated in Fig. 3-3. This figure is separated into two broad parts: the functional objectives that are accomplished *inside the controller case* and the balance of the process control loop that is, of course, external to the controller case.

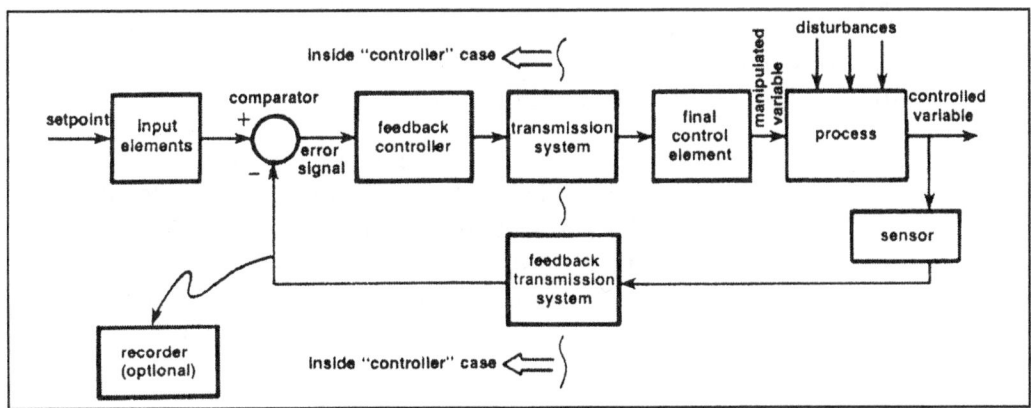

Figure 3-3. The Functional Layout of a Feedback Loop

Provision must be made for either the operator or some hardware to provide a set point to the control loop. This set point is the desired value of the controlled variable and will have the same dimensions as the controlled variable; for example, if the controlled variable is gallons per minute then the set point also will be gallons per minute. The input elements provide a functional conversion of the input set point signal into the operating units of the controller, for example, millivolts (mV), milliamps (mA), or pounds per square inch gage (psig) air pressure.

The controlled variable is actually measured by a sensor, and the measured value of the controlled variable is transmitted back to the controller case. Inside the controller case is the *comparator*. This functionally important device represents the controlled variable itself by actually comparing—taking the algebraic difference between—the value of the set point (after conversion by the input elements) and the value of the variable transmitted back into the controller case. The comparator, or *error detector*, is common to all feedback control systems. Note that this is negative feedback control, that is, the signal fed back to the comparator is subtracted from the signal that represents the set point. All applied feedback control is negative feedback control (positive feedback control is inherently unstable).

The error signal, that is, the output of the comparator, becomes the input to the feedback controller. Proper, formal process instrumentation terminology refers to this error signal as the *actuating error signal* or the *system deviation* (see the glossary in Appendix B). Based on the error signal, the controller calculates a signal to the final control element, typically a control valve, and this in turn controls the manipulated variable input to the process.

Also shown with the feedback control loop in Fig. 3-3 is a recorder (for the controlled variable), which is optional. In practice, many people refer to all of the items contained within the controller case as the feedback controller. While common, this broad usage is not necessarily precise and will not be used in this book. Instead, it is more desirable to make functional distinctions among the various operations performed around the feedback control loop and within the controller case.

Because electronics and microprocessors and software have all had a major impact on process control equipment, the loop as we have described it here has become smaller and smaller, shrinking into a single electronic unit or a few lines of code. But this in itself leads to a "black magic" view of feedback control. Thus, all the component parts are shown here in their stand-alone and "clunkier" detail because as a simple matter they are thus easier to visualize for the beginning student.

3-4. Dynamic Components

The various blocks of a feedback control loop have different types of dynamic behavior, and it is most important that you gain insight into these *process dynamics*. Although we will treat this in more detail in Unit 8, an introduction to the subject is appropriate here.

Many of the individual components of the process control loop have no time-dependent behavior, that is, there is no *lag* or *lead* in their operation. When the input to the component changes, for all practical purposes the output changes instantaneously:

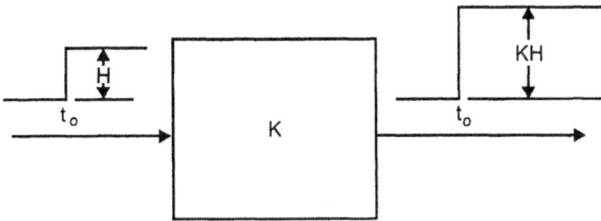

The type of component illustrated will be referred to as a *nondynamic* component. In effect, in a mathematical sense it is algebraic in nature. The output changes instantaneously (for all practical purposes) when the input changes. In effect, the output is always proportional to the input, and this proportionality constant will be referred to as the *gain* K of the component. K is dimensionless. Many individual components illustrate "dynamic" characteristics. Typically, their output will lag or lead any input; lag is far, far more common. When such time-dependent behavior is encountered, it appears like this:

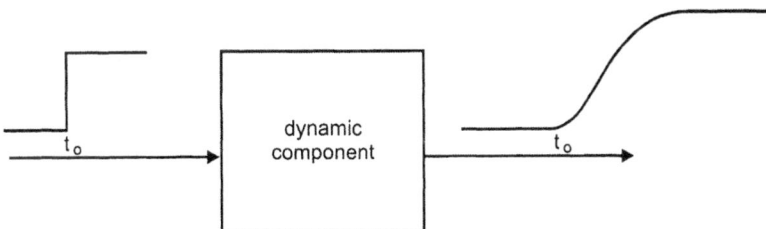

This type of dynamic response is typically encountered in the process itself, and to a lesser extent in the control valve and in the sensor. The specific mathematical form of these dynamic lags will be discussed in more detail in Units 3-5, 3-6, and 8.

3-5.   Mathematical Model of a Loop

In Fig. 2-4 and Fig. 3-3 you looked at a functional layout of a feedback control loop. This is an appropriate time for you to understand this functional layout with some mathematical and dynamic insight into its behavior. The feedback control loop might be as illustrated in Fig. 3-4.

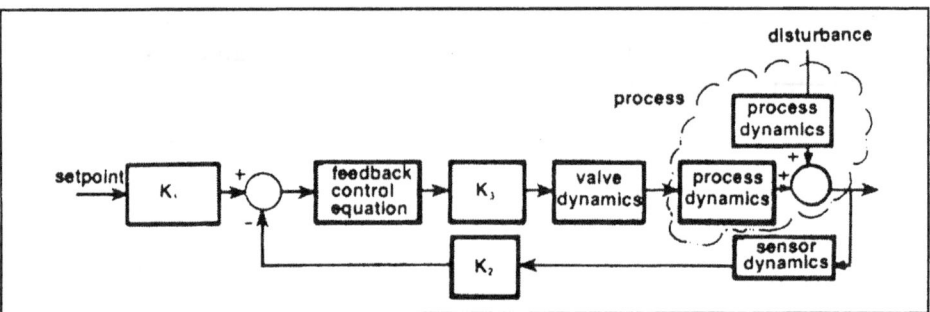

Figure 3-4. The Mathematical Layout of a Feedback Loop

Note that the input elements are represented as a nondynamic component, that is, when the set point changes the signal to the comparator tends to change relatively instantaneously. The two transmission systems are also shown to be nondynamic. If they are well designed, they should not have significant time lags associated with them. (Sometimes, in older pneumatic transmission systems, this becomes a problem, but this will be discussed later in Unit 4.)

To a lesser extent, both the sensor and the valve will have their own dynamics, but in the typical case the dynamics (the lag) of these individual components will be much less than that of the process itself. All process loops are functionally the same, and, in general, they follow the layouts we presented in this book. Process dynamics vary significantly from one individual loop to another, and the practitioner must gain some appreciation and insight into the dynamics of an individual loop in order to design, install, and tune the loop to provide quality control.

Quite often, students get distressed about the need to mathematically or quantitatively analyze dynamic performance. This frustration is understandable, but it does not eliminate the necessity for dealing with process dynamics. Process control is obviously needed only in situations that are changing, that is, if nothing is changing you do not need any control. Things that are changing are doing so with respect to time, and it is important to understand their dynamic behavior. As a result, to understand process control one must appreciate and understand process dynamic behavior.

As illustrated in Fig. 3-4, the process itself has dynamic lag all its own, and when disturbances enter the process they will produce an effect on the controlled variable that is dynamic in nature. The same thing can be said of the manipulated flow as it enters the process.

3-6. Mathematical Notation (For Those Who Want It)

There are no universally accepted standards for the mathematical notation used in process control presentations, and as a result practice varies. For those to whom the mathematics is important, we present the notation used in this book as well as its implications here. For those who are mathematically impaired, do not feel you must enjoy all these details to see the big picture that unfolds throughout the subsequent units of this book. In this book the following conventions will be observed. For the generalized mathematical representation of an element or a system, we will use $G(p)$ for the representation in the time domain. p is the Heaviside operator that indicates taking the derivative with respect to time, d/dt. $1/p$ implies $\int ... \, dt$. In its simplest form, $G(p)$ may imply simple algebraic multiplication, and in more complicated (time-dependent) situations, it may represent a differential equation.

The following is a first-order differential equation (first-order lag) example:

$$G(p) = \frac{K}{(1 + \tau p)}$$

where K is the gain, τ is a time constant, and p is the Heaviside operator. In block diagram notation, this may be represented as

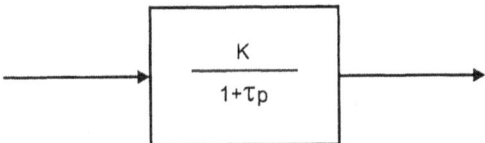

It is inconvenient to express dead times θ in the time domain, and thus most presentations of process control shift to the Laplace domain. This domain employs the Laplace transform of time domain functions. In simple situations, the symbol p simply becomes s. Our earlier example simply becomes

$$G(s) = \frac{1}{(1 + \tau s)}$$

or in block diagram form:

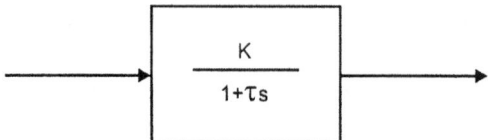

In some cases, where we are using engineering units for the input and output signals from an element (or block), we use the term *sensitivity* instead of *gain*. In this case, the symbol S is used instead of K. For example,

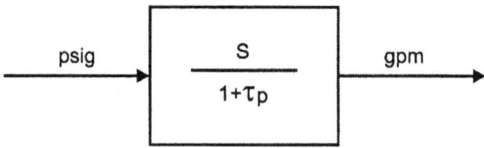

where S has the units gpm/psig.

Do not be distressed. It will start to make sense quickly.

EXERCISES

3-1.   *Fig. 2-2 shows a sketch of a simple home heating system. Develop a functional layout of the basic feedback control loop of this system.*

3-2.   *Given a simple level control system as shown here, develop a functional layout of the basic feedback control loop of this system:*

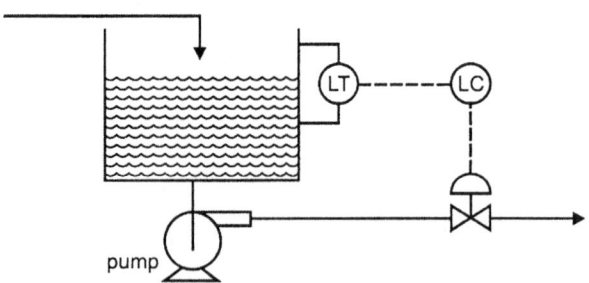

3-3.   *A temperature sensor has a range of 0-200°F and transmits a 3-15 psig air signal based on the measurement. What is its sensitivity?*

3-4. *The input elements of an electronic controller accept a set point of 0 to 60 gpm and produce a 4 to 20 mA signal to the comparator. What is the sensitivity of these input elements?*

3-5. *Estimate whether or not the dynamic characteristics of the following items of hardware would be significant in the total dynamics of a feedback control loop:*

- *the input elements in a controller*
- *an electronic transmission system*
- *a large, pneumatically operated control valve*
- *a 2,000 ft. pneumatic transmission system*
- *a pneumatic controller*
- *an electronic controller*
- *a bare thermocouple*
- *an orifice meter*
- *a chromatograph*
- *a thermocouple encased in a heavy thermowell*

To answer this question, you must obviously make guesses about whether or not the listed items are "relatively fast" or "relatively slow." Thus, there can be no certainty as to what is "right" or "wrong" in your answer. When you ask a relative question, then you must ask "Compared to what?" In process control the answer is "Compared to the other elements around the loop." This leads to an appreciation of relative speed, and that is the important point at this stage of your education.

3-6. *In general terms as well as for typical industrial applications rank the following processes as to whether you would consider them to be (a) responding very rapidly to input changes, (b) responding at a moderate rate, or (c) responding very slowly:*

- *liquid flow in a line*
- *gaseous flow in a line*
- *liquid level in a small tank*
- *liquid pressure in an enclosed tank*
- *gaseous pressure in a large tank*
- *composition in a large distillation column*
- *temperature in a liquid-filled tank*
- *your coworkers at 5:00 p.m.*

The comment about making guesses at the end of Exercise 3-5 applies here as well.

Unit 4:
Sensors and
Transmission Systems

UNIT 4

Sensors and Transmission Systems

The quality of performance of a feedback control system is directly dependent on the quality of measurement of the controlled variable. This measured value must be transmitted to the controller in a timely fashion so that corrective action may be calculated. Then the controller output must be transmitted to the control valve. The purpose of this unit is to study the operation of these measuring and transmission elements.

Learning Objectives — When you have completed this unit, you should:

A. understand more fully the role and functioning of the sensors in a feedback control loop

B. be able to define accuracy, precision, and sensitivity

C. have insight into what constitutes good dynamic behavior in a sensor

D. appreciate the characteristics of transmission systems

4-1.   The Sensor and the Transmitter

One of the most critical problems in designing and installing a feedback process control system is specifying the sensing device that will obtain a measurement of the controlled variable. This measuring device not only provides a measurement of the controlled variable; it also causes a *change of variable*. The change of variable takes place because the controlled variable itself is not the actual signal that is transmitted back to the comparator. The *transmitter* produces an output signal whose steady-state value has a predetermined relationship to the controlled variable.

A transmitter is not required from the standpoint of either measurement or control. The transmitter is an operating device to make the measured value of the controlled variable available in a more convenient and/or a more centralized location, for example, in a remote control room. The transmitter output signal is generally in a standardized form, such as 3 to15 psig air pressure or 4 to 20 mA DC. From a hardware standpoint, the measurement function and the transmitter function are often both incorporated into a single device.

The main controlled variables in process control systems are as follows, in descending order of the frequency of their occurrence: *temperature, pressure, flow rate, composition,* and *liquid level.* Some of these variables, such as pressure, can be measured relatively directly, while others, such as temperature, can be measured only indirectly.

One additional device to be described is the *transducer.* This is a device that receives information in one physical form, modifies the information or its form or both, and sends an output signal. Depending upon the application involved, a transducer in a functional sense can be a primary measuring element (a sensor), a transmitter, a relay, a converter, or some other device.

You can appreciate that many of these terms—sensor, transmitter, converter, transducer—have broad and sometimes overlapping meanings, and quite often the particular installation of a specific piece of hardware will dictate the appropriate descriptive term. Confusion can be minimized if the practitioner focuses on the function involved rather than the term.

4-2.   Sensor Dynamics

It is important to understand sensor dynamics. Quite often the speed of response of the primary measuring element is one of the most important factors affecting the operation of a feedback control loop. Obviously, process control is continuous and dynamic, and the rate at which the controller can detect changes in the controlled variable will be critical to the overall operation of the system.

To gain some understanding of sensor dynamics, refer to Fig. 4-1, which that shows a bare bulb-type expansion thermometer. For analysis purposes, suppose that we immerse this bare bulb into an agitated constant temperature bath as shown in Fig. 4-1.

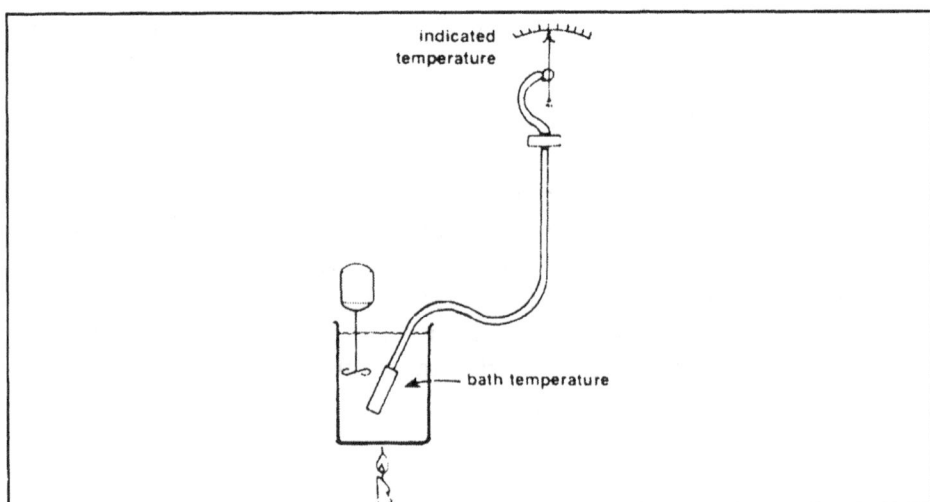

Figure 4-1. Thermometer Experiment

The bare bulb will make the transition from ambient temperature to the temperature of the bath, and the thermometer needle might rise, as shown in Fig. 4-2. The curve shown in Fig. 4-2 is exponential and approaches the bath temperature gradually.

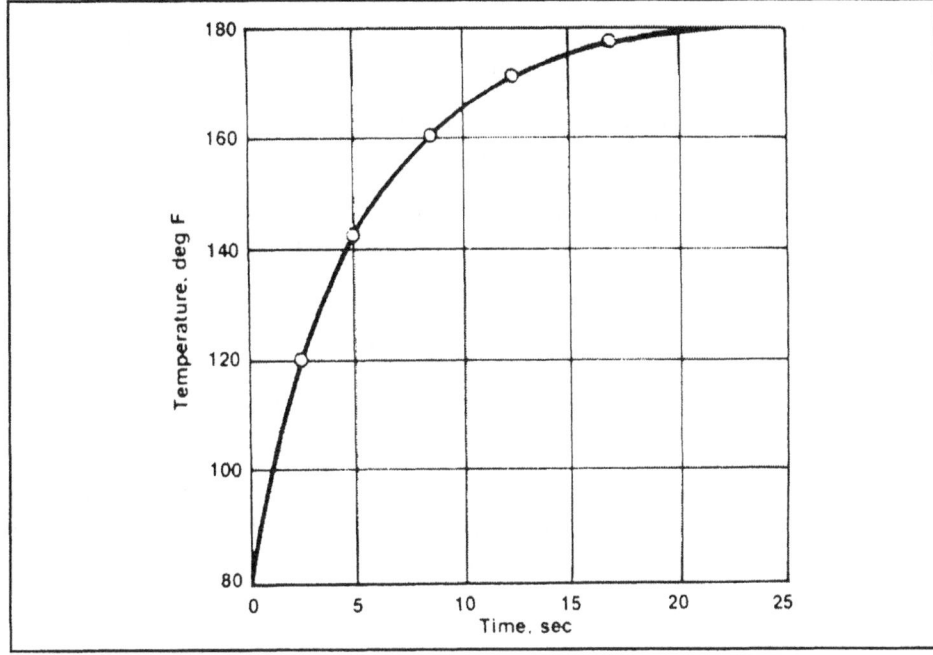

Figure 4-2. Response of Bare Bulb Thermometer

A curve like the one shown in Fig. 4-2 is referred to as a *response curve,* and it gives experimental insight into the dynamics of this particular measuring device.

Unit 8 will cover process dynamics in some detail. For the moment, we will introduce the study of process dynamics by defining a term that is used to characterize the dynamic behavior of a response curve like the one shown in Fig. 4-2. This term is the *time constant* of the bulb, and it refers to the time needed for the response curve to reach 63.2 percent of its final value. In Fig. 4-2, the time constant for the bulb is approximately five seconds. The physical meaning of the term *time constant* will be explored in Unit 8, but for now we will use the term to make some quantitative comments about sensor dynamics.

It is the signal shown in Fig. 4-2 that is typically available for transmission, and we see that some lag is introduced into the feedback loop by the sensor. It is desirable to minimize such lag as much as possible. Fast sensors make it possible for the controller to function quickly. Sensors with large time constants are slow and degrade the overall operation of the feedback loop. You should consider the dynamic characteristics of sensors when you are selecting and installing them.

4-3. Selection of Sensing Devices

Many questions must be considered before selecting a specific sensor for a particular loop. There are no hard and fast rules for making such decisions, but there are a number of factors that must be considered:

    a.   What is the normal range of the controlled variable? Are there extremes to this range?

    b.   What accuracy, precision, and sensitivity are necessary? (These terms are defined in detail in the next section.)

    c.   What sensor dynamics are needed and available?

    d.   What reliability is required?

    e.   What are the costs involved—not simply the purchase cost but also the installation and operating costs?

    f.   Are there special installation problems, for example, corrosive fluids, explosive mixtures, size and shape constraints, remote transmission questions?

With such a long list of important factors to keep in mind, selecting particular sensors for specific installations is very complex. As a result, whole books have been devoted to such individual subjects as how to measure temperature, how to measure flow rate, how to measure composition, as well as other common process variables. To study these subjects, you may want to refer to these books. ISA publishes many of them. We will not explore in further detail the particular details of sensor selection for now. Instead, we will focus on them only to the extent that they affect the basic performance and dynamics of the control loop.

4-4. Accuracy and Precision

The *accuracy* of a measurement refers to the closeness with which the measurement approaches the true value of the variable being measured. *Precision* refers to the magnitude of random error. Whenever a measurement is made, sources of precision or random error will add a component to the result that is unknown. However, with repeated measurements made under the same conditions, this result changes in a random fashion. (Interestingly, precision is not a defined term in the ANSI/ISA standard S51.1-1979 (R 1993) on "Process Instrumentation Terminology," but it is used routinely by many measurement professionals.)

A more common term related to precision is *repeatability*, which is discussed in the next section. In matters of process control, precision and repeatability are more important than accuracy. To put this differently, it is normally more desirable to measure a variable precisely than it is to have a high degree of absolute accuracy. The distinction between these two properties of measurement is shown in Fig. 4-3.

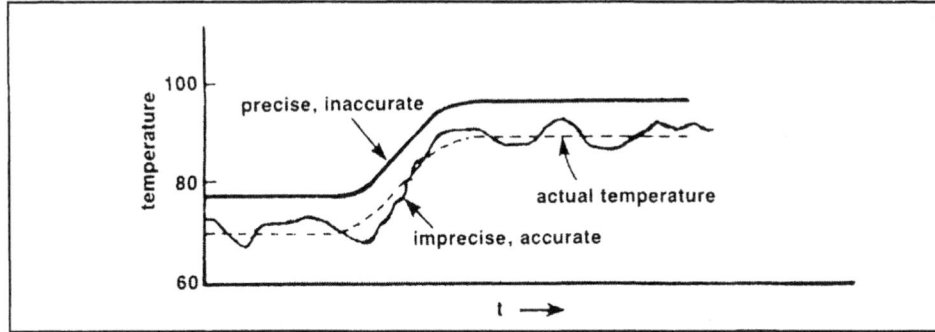

Figure 4-3. Accuracy and Precision in a Temperature Measurement

The dashed curved line in the figure indicates the actual temperature of a fluid. The upper measurement or solid line illustrates a precise but inaccurate instrument. The lower measurement shows the measurement given by an imprecise but more accurate instrument. The first instrument (the upper measurement) has the greater "error."

"Target" or bull's-eye illustrations, such as those shown in the figure for Exercise 4-7, are quite useful in understanding the terms accuracy and precision. The student should find it quite helpful to study this example and its answer, which is given in Appendix C.

Practitioners make a distinction between two types of accuracy: *static* or *steady-state accuracy* and *dynamic accuracy*. Static accuracy is the closeness of the approach to the true value of the variable when that true value is constant. Dynamic accuracy, on the other hand, is the closeness of approach of the measurement when the true value is changing. These terms are illustrated in Fig. 4-4.

The numerical value of the dynamic accuracy will depend on the nature of the dynamic change made by the true value of the variable being measured. The properties of the measuring system itself will also have an effect. For process control systems, a practical specification of the time variable is *ramp forcing*, like that shown in Fig. 4-4, and one practical designation of dynamic accuracy is the *dynamic error* that results from ramp forcing.

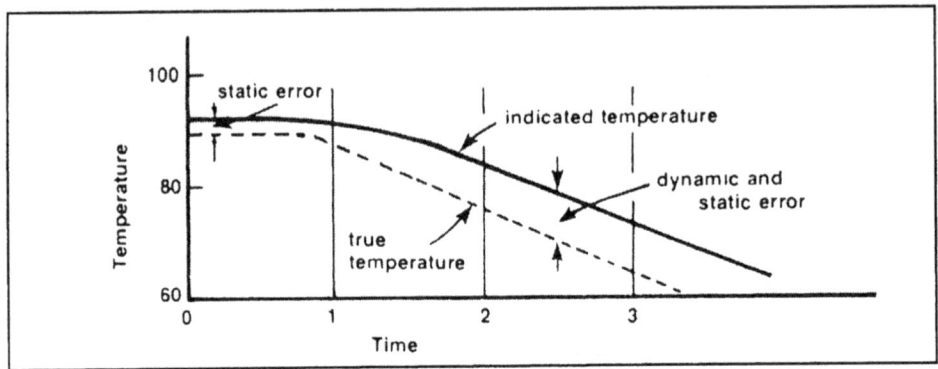

Figure 4-4. Dynamic and Static Error

4-5.   Sensitivity, Repeatability, and Reproducibility

The *sensitivity* of a measuring device is defined by the ratio of the output signal change to the change in the measured variable. Sensitivity has the engineering units of the output signal divided by the engineering units of the input signal. Clearly, the greater the output signal change for a given input change, the greater the sensitivity of the measuring element. Sensitivity is a steady-state ratio, and, in effect, it is a steady-state characteristic of the element.

There is another kind of sensitivity that is very important in measuring systems. This sensitivity is defined as the smallest change in the measured variable that will produce a change in the output signal from the sensing element. In many physical systems, especially those that contain levers, linkages, and mechanical parts, these moving parts have a tendency to stick and to have some *free play.* The result is that small input signals may not produce any detectable output signal. Well-designed and well-constructed instruments are needed so that sensitivity will be high, and so the control system has the ability to respond to small changes in the controlled variable.

Repeatability is the closeness of agreement among a number of consecutive measurements of the output for the same value of the input under the same operating conditions, approaching from the same direction, for full-range traverses. The degree of repeatability achieved between consecutive measurements is usually measured as *nonrepeatability* and is specified as a percentage of full scale or a percentage of span (span being the difference between the full-scale and the zero-scale value). Repeatability thus measured does not include hysteresis.

Reproducibility is the closeness of agreement among repeated measurements of the output for the same value of input made under the same operating conditions over a period of time, approaching from both directions. Reproducibility includes hysteresis, dead band, drift, and repeatability, which are defined in the glossary, Appendix B.

4-6. Rangeability and Turndown

Rangeability refers to the minimum and maximum measurable values of the process variable being measured. For example, consider a flowmeter where the maximum measurable flow is 100 gpm and the minimum measurable flow is 10 gpm. The rangeability of the flowmeter is 10 percent to 100 percent.

Turndown is another way to measure rangeability, but in which the information is expressed differently. Turndown is the ratio of the maximum measurable flow to the minimum measurable flow. For the flowmeter described in the previous paragraph, the turndown is 10 to 1.

4-7. Measurement Uncertainty Analysis

Measurement uncertainty analysis is a numerical, objective method for defining the potential error that exists in all data. It is crucial to have knowledge of the uncertainty in any measurement to understand either the state of a process or its performance.

In measurement uncertainty analysis, errors are considered to be either random errors (sometimes called precision errors) or bias errors. Random error adds a component to a measurement that is unknown; but, with repeated measurements, it will change in a random fashion. Such random errors are drawn from a distribution of error that is Gaussian normal.

Uncertainties in measurements may be due to bias, offset, or scale-shift errors; span, gain, or sensitivity errors; errors of curvature or lack of conformity such as nonlinearities; randomness such as that caused by measurement noise or vibrations; and, last but not least, the inability to resolve the measurement to better than a certain amount, that is, precision.

It is not sufficient to focus only on reducing precision errors. It is also necessary to understand and reduce systematic errors or bias. Systematic errors or bias are constant in the sense that they affect every measurement of a variable by the same amount. They are not observable directly in the measurements themselves.

Measurement uncertainty analysis is a field that merits study by serious process control students. It gives clearer insight into a poorly appreciated—but most important—phenomenon.

4-8.   Transmission Systems

When the sensor measures the controlled variable, the measured value must be transmitted in some way to the controller. This may be a distance of several feet or several thousand feet. In a similar fashion, it is necessary to get information from the controller to the final control element, such as a control valve.

For years, many process control transmission systems used pneumatic tubing to transmit information as an air pressure signal. Very few of these pneumatic transmission systems are being installed today; but a very large number of them are installed in existing plants. These pneumatic systems introduce a time lag into the process dynamics of the control loop and thus influence loop performance—sometimes to serious disadvantage. For all of these reasons, some discussion of pneumatic transmission systems is warranted here.

The most common transmission medium is copper wire, either in the form of twisted pairs or in the form of coaxial cable. Twisted pairs have been used extensively for years and are the most common transmission system. Wireless technologies can also be useful, especially in mobile or temporary installations. Fiber optic cable can also be used, and its installation is becoming increasingly common. But it is very expensive, and it is difficult to make multiple splices.

Many transmission systems are available and in use, and we will review some of the more common ones here. Table 4-1 lists the common transmission systems in use today (except for digital buses). We will concentrate on the 3-15 psig analog pneumatic type and the 4 to 20 mA DC type.

Type	Medium	Values
Binary (on-off)	Electricity: alternating current direct current	0 to 120 volts 0 to 24, 48, or 125 volts
	Pneumatic	0 to 25, 35, 100 pounds per square inch gage (psig) (170, 240, 700 kilopascals)
	Hydraulic	0 to 3000 psig (20,000 kilopascals)
Analog (modulating)	Electricity: direct current	−10 to +10 volts, 1 to 5 volts, 4 to 20 milliamperes, or 10 to 50 milliamperes
	Pneumatic	3 to 15 psig, 6 to 30 psig (20 to 100 kilopascals, 40 to 200 kilopascals)
	Hydraulic	0 to 3000 psig (0 to 20,000 kilopascals)

Table 4-1. Typical Transmission System Signal Ranges (Ref. 7)

4-9. Pneumatic Transmission

Pneumatic transmission may be used for distances up to several hundred feet. In the feedback transmission system, the controlled variable is measured and converted to an air pressure, and a transmitter, in effect, sends this air pressure signal through a single tube to a receiver where it is transduced to a position or force for use within the controller. A typical pneumatic transmission system is shown in Fig. 4-5. The controlled variable is sensed and converted to an air pressure, and the measured pressure is often used as a pilot signal to an amplifier. (An amplifying pilot is often employed in pneumatic transmission systems to increase the air flow capacity of the transmitter and to linearize the output signal.)

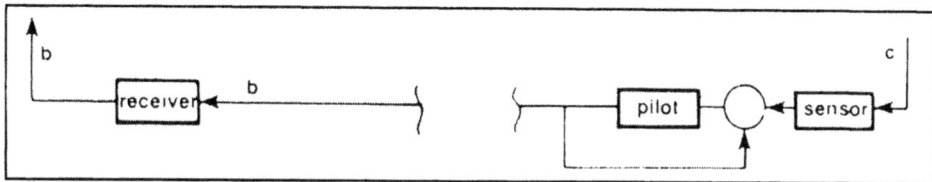

Figure 4-5. A Pneumatic Transmission

The connecting tube carries the transmitted pressure to a receiver (located in the controller case). This tube is almost always one-quarter inch in outside diameter and may be copper, aluminum, or plastic. The receiver is simply a pressure-gage element, and the transmitted air pressure is converted into the movement of a bellows or diaphragm. Thus, pressure is transduced into a position or force that is used within the controller.

One of the most serious concerns in pneumatic transmission is the problem of transmitting a signal over significant distances because there is a lag associated with the transmission of the pressure signal through the long connecting tube. The tube has volume distributed along its length, and there is resistance to fluid flow through this tube. If a step change in input signal of 3 to 15 psi is made to a pneumatic transmission tube, then a typical output signal would appear as shown in Fig. 4-6.

As distances increase, the speed of response of pneumatic transmission systems becomes a problem, and alternate solutions must be found. Very few pneumatic transmission systems are being installed today, but there is a very, very large number of these systems still in use in industry.

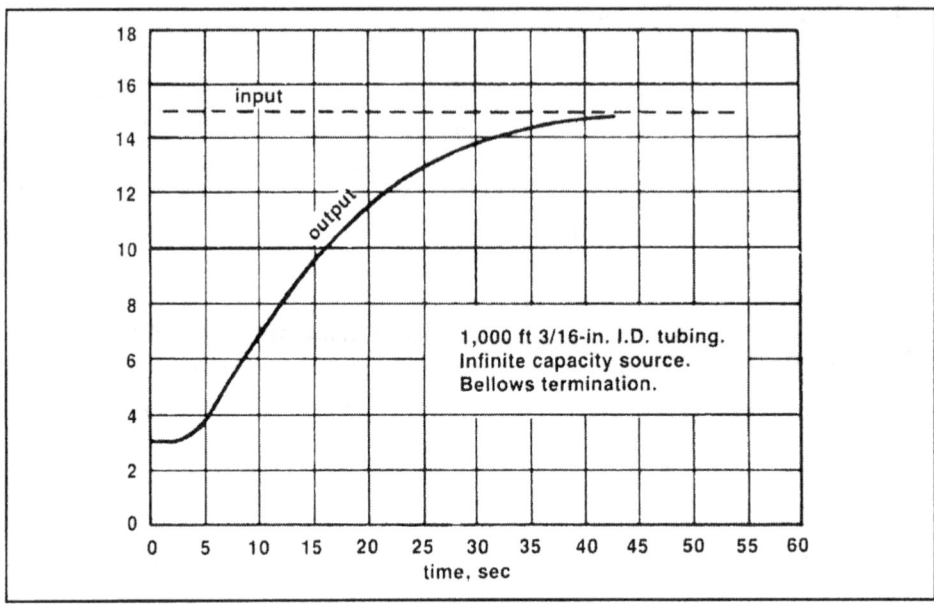

Figure 4-6. Input and Output of a Pneumatic Transmission System

4-10. Electrical Transmission Systems

The most common transmission system electrical signal is 4 to 20 mA. A twisted pair of copper wires is used to form a DC current loop. Current is preferred over voltage because it is more immune to transmission line length and its associated electrical resistance, and it requires only two wires. There is an individual twisted pair of wires for each signal, and for a fluid process plant this rapidly creates the need for literally thousands of such twisted pairs. They are very expensive to install, and the installation may cost several times more than the process control equipment itself.

Field wiring must also provide for the galvanic isolation of signals from the process equipment in order to prevent ground loops. Field wiring is also often installed in hazardous locations that require explosion-proof or intrinsically safe media. It is possible to consider using a sensor output directly as a signal, for example, using the mV output of a thermocouple as a direct signal. Problems with noise, signal weakness, and the like will normally lead you to use signal conditioning in the form of a *transmitter* to assist transmission.

There are two types of transmission systems—the two-wire system and the four-wire system. In the two-wire system, the current that powers the conditioner/transmitter also carries the signal. The four-wire type has separate wires for power. See Fig. 4-7.

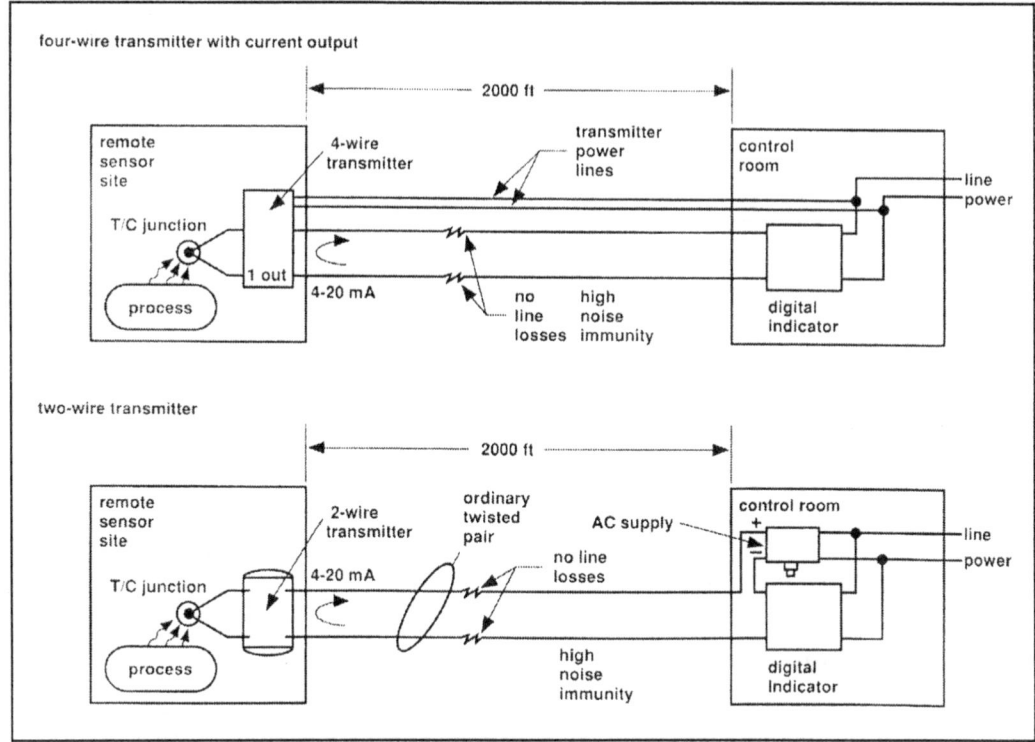

Figure 4-7. Two-Wire and Four-Wire Operations (Ref. 2)

4-11. Multiplexing

With so many signals that must be conditioned in a similar way and transmitted to the same place, for example, to the control room, there is a need to devise a system so that various signals can time-share portions of the transmission system. You can do this by switching the system from one signal to another. This is called time-share *multiplexing*, and the switching is called *scanning*. This is illustrated in Fig. 4-8. The scanning rate can be fifty or more points per second. When a single pair of twisted wires is used to carry numerous signals in this manner, it is called a *data highway*. Multiplexing leads us to the *digital fieldbus* concept.

4-12. Digital Fieldbus

As transmission systems become more complex and move more and more into the digital world, it has become obvious that there is a need for a standard digital communication system for use in process control applications. The digital fieldbus is the term used for the digital replacement of the 4 to 20 mA DC systems used to communicate between the elements of a process control system (see Ref. 3). Digital fieldbus will be discussed in more detail in Section 15-2.

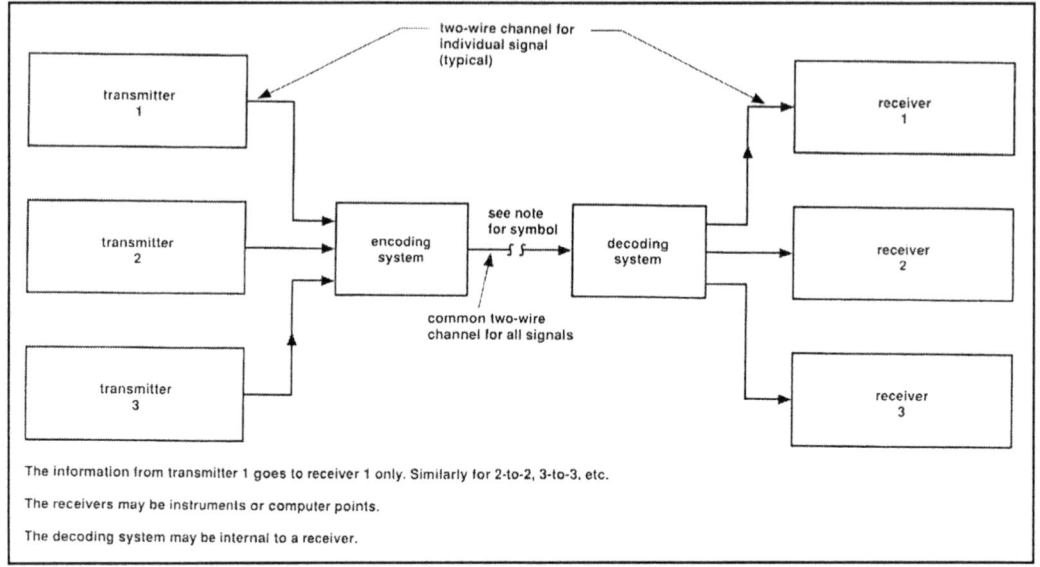

The information from transmitter 1 goes to receiver 1 only. Similarly for 2-to-2, 3-to-3, etc.

The receivers may be instruments or computer points.

The decoding system may be internal to a receiver.

Figure 4-8. Basic Multiplexing (Ref. 7)

4-13. Smart Sensors

The term *smart sensors* refers to those devices that have microprocessors built into their internal operation that control the devices. The signal sensed by the measuring device is brought into the transmitter and is immediately converted into a digital signal. All internal signal conditioning and processing is done digitally. Tremendous gains are made in accuracy, linearity, and versatility. The output of the transmitter can be in digital format, or it can be converted to a 4 to 20 mA signal.

In addition to the advantages just cited, smart sensors can themselves be *programmed* by the computer that controls the loop where the sensor is used. The options for improvement are extensive and are just beginning to be fully exploited by the hardware vendors.

REFERENCES

1. J. Bryzek and R. Grace, "Silicon's Synthesis: Sensors to Systems," *InTech*, vol. 37, no. 1 (January 1990), p. 40.

2. Donald R. Carlson, "Temperature Measurement in Process Control," *InTech*, vol. 37, no. 10 (October 1990), p. 26.

3. Richard H. Caro, "SP50 Chair Explains How Buses Relate," *InTech*, vol. 43, no. 11 (November 1996), pp. 18-19 (last of a four-part series in 1996).

4. Ronald H. Dieck, *Measurement Uncertainty: Methods and Applications*, 2d ed. (ISA, 1997).

5. Albert A. Gunkler and John W. Bernard, *Computer Control Strategies for the Fluid Process Industries* (ISA, 1990).

6. Clifford W. Lewis, "The Smart Future of Two-wire Temperature Transmitters," *InTech*, vol. 37, no. 2 (February 1990), p. 42.

7. George Platt, *Process Control: A Primer for the Nonspecialist and the Newcomer*, 2d ed. (ISA, 1998).

8. P. Chapman, *Smart Sensors* (ISA, 1996).

9. L. M. Thompson, *Industrial Data Communications: Fundamentals and Applications*, 2d ed. (ISA, 1997).

10. D. T. Miklovic, *Real-Time Control Networks* (ISA, 1993).

11. J. C. Huber, *Industrial Fiber Optic Networks* (ISA, 1995).

EXERCISES

4-1. *Our entire presentation of sensors and transmission systems in this unit was structured around the idea of feeding a signal back to the comparator. What would be involved (how would the hardware be different) if the process variable being measured were not a controlled quantity but was instead simply being sent to a central control room for indication and recording?*

4-2. *Refer to the immersion of the bare bulb in the bath of Fig. 4-1. How would the process response curve be different if the thermometer bulb were not bare but instead encased in a protective thermowell?*

4-3. *A process chromatograph is used as a sensor to measure the composition of a liquid stream. It operates on a discrete basis, that is, it takes a sample and analyzes it and establishes a signal for feedback purposes. It then takes a new sample and starts its cycle over again. What effect does this mode of operation have on loop operation as compared to a continuous sensor such as, for example, a thermocouple or an orifice meter?*

4-4. *In Section 4-4, we stated that sensor repeatability is more important than accuracy. How can this possibly be true?*

4-5. *How far can you transmit a pneumatic signal if you can tolerate a time-constant lag of three seconds and a dead time of less than one second?*

4-6. *In a feedback control loop there are two transmission systems: one to feed back the signal that indicates the controlled variable and one to send the controller output to the control valve. Are the process dynamic characteristics of one of these systems more critical than the other?*

4-7. *Consider the four parts of the figure for this exercise. Characterize or describe the four patterns shown in view of the discussion of measurement uncertainty in Section 4-7 (see Ref. 4).*

(a) tight arrows, centered

(b) spread arrows, centered

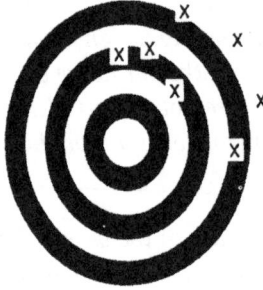

(c) spread arrows, off center

(d) tight arrows, off center

Unit 5:
Typical Measurements

UNIT 5

Typical Measurements

For process control to function, you must be able to sense/measure the important variables of the process. This unit will provide a brief survey of some of the more common sensors used to measure process variables.

Learning Objectives — When you have completed this unit, you should:

A. be able to explain how pressure transmitters work

B. be able to describe the operating principle of an orifice flowmeter

C. be able to describe at least two common-level sensors

D. be able to describe how a thermocouple operates

E. be able to describe how a pH meter and a gas chromatograph operate

5-1. Pressure Sensors/Transmitters

Pressure is one of the most common and most important measurements associated with process control. Most of us are familiar with pressure measurement devices such as manometers or dial-pressure gages. These devices assume that a person will be there to read them. To be useful in process control, a pressure-measuring device must have a pressure transmitter that will produce an output signal for transmission, such as an electric current that is proportional to the pressure being measured. Typically, a transmitter produces an output in the form of a 4–20 mA signal, is rugged, and is rated for use in electrically hazardous areas.

Pressure is defined as force per unit area; some common units are pounds per square inch (psi), inches of water, inches of mercury, atmospheres, bars, and torrs. Pressure is always measured against one of three common references. If the reference is a vacuum, the measured pressure is called *absolute pressure*. If the reference is local ambient pressure, the measured pressure is called *gage pressure*. If the reference is user-supplied, the measured pressure is called differential pressure. Fig. 5-1 illustrates these three reference types; P_2 is the reference pressure.

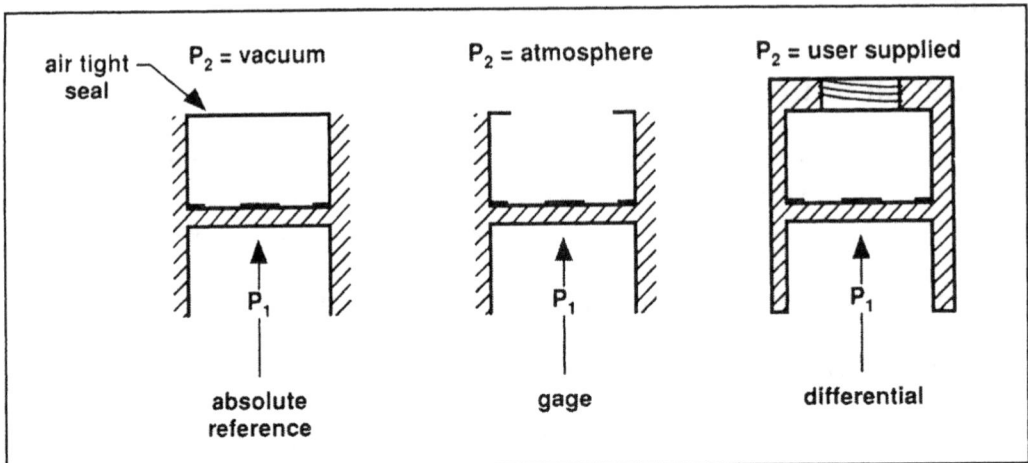

Figure 5-1. The Three Types of Pressure References

One of the most common ways to measure pressure in process control applications is to use a strain gage sensor like the one shown in Fig. 5-2. The pressure produces a strain on a diaphragm that causes a displacement of the diaphragm. This displacement is then measured and translated into a pressure measurement. In Fig. 5-2, the strain gage is an electrical resistor whose resistance varies as a function of how much it is bent. Four of these strain gages are bonded onto a metal diaphragm. When pressure is applied, two of these four strain gages will be in compression (their resistance increases). The four strain gages are connected into a Wheatstone bridge circuit to yield an electrical signal that is proportional to the strain, displacement, or pressure.

The transmitter's housing contains not only the strain gage sensor but also a signal conditioner, a pressure input port, and an electrical connector. The signal conditioner includes all of the electronics needed to compensate, linearize, adjust, and amplify the signal. The pressure source is connected to the pressure input port, and the electrical connector is used to connect the output signal.

5-2.   Pressure Transmitter Technologies

In the previous section, we looked at one common type of pressure transmitter. This section will review some of the various technologies that are employed by the many hardware vendors that supply pressure transmitters.

Strain Gage

This technology was discussed in the previous section. There are many different ways to make a strain gage and to bond it to the diaphragm.

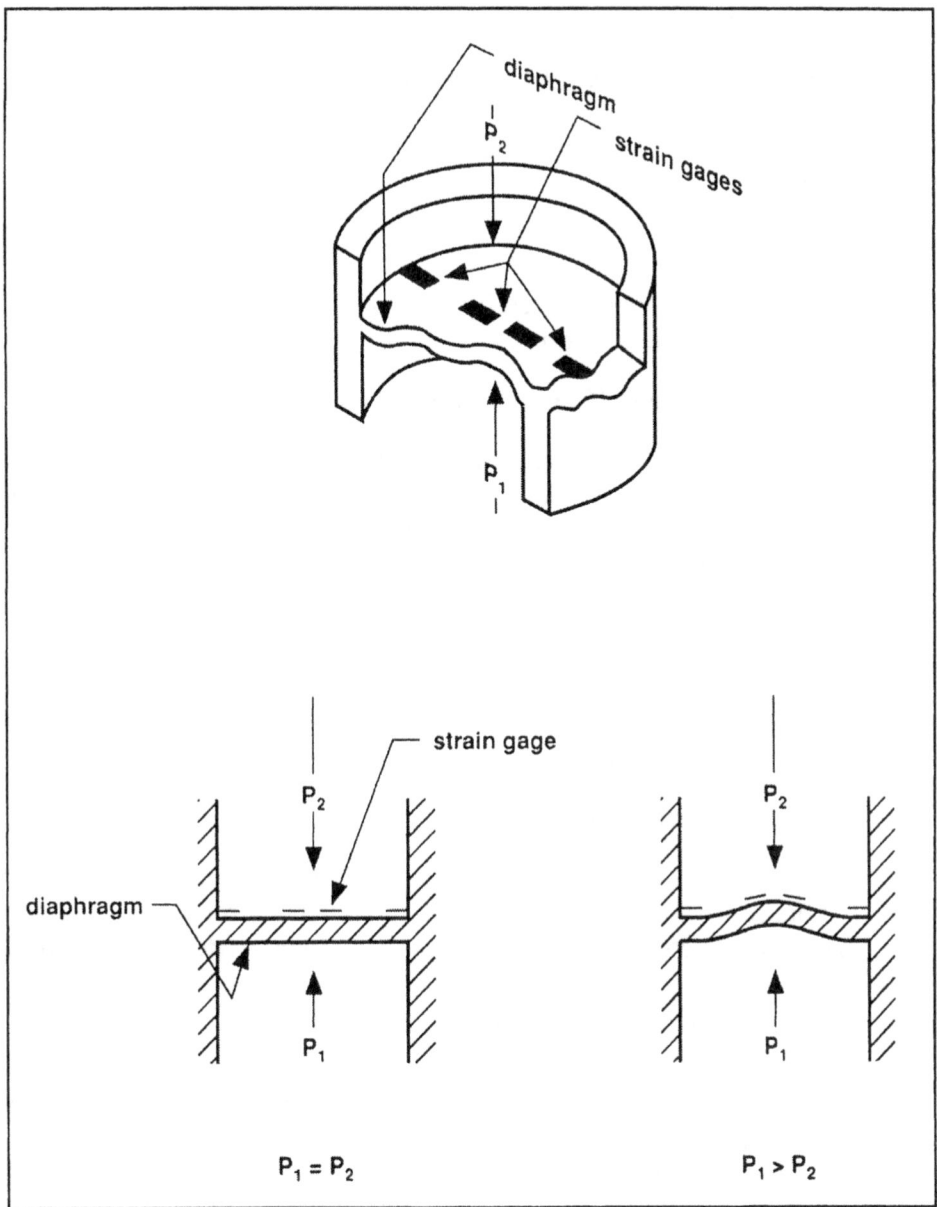

Figure 5-2. A Strain Gage Sensor

Vibrating Wire

A thin wire inside the sensor is in tension, with one end fixed and one end attached to a diaphragm. As pressure moves the diaphragm, tension on the wire changes, and its resonant vibration frequency changes. These changes in the frequency of the vibrating wire are used to measure pressure.

Capacitance

Placing a diaphragm between two stationary electrodes forms two capacitors. Pressure changes move the diaphragm, changing the relative capacitance of both capacitors and, thus, providing a measure of the pressure.

Thin Film

This is a strain gage formed by the vapor deposition of a metal onto an oxide-coated diaphragm.

Linear Variable Differential Transformer

This is a variable-reluctance transmitter in which the ferrite core is replaced by a coil with an AC voltage exciter. Pressure moves this exciter, and the induced voltage in each coil is different because of the displacement.

Fiber Optic

An optical fiber is used to transmit light that is reflected off a diaphragm to a receiving fiber. As pressure moves the diaphragm, the light received is changed, thus indicating the change in pressure.

"Smart" Transmitters

More and more transmitters are employing microprocessors, especially for signal conditioning. These devices can even be programmed remotely. They can be coupled with any of the technologies described in this section.

5-3.   Flow Sensors/Transmitters

More than 75 percent of all transmitter sales are for measuring flow. As you might expect with such a huge market, flowmeters come in all imaginable types, shapes, and sizes. A brief survey of the most common types of flowmeters will help you make sense of this variety.

All flow-measurement devices operate using one of two approaches—either energy extractive or energy additive. Fig. 5-3 and Fig. 5-4 show the various flowmeters available within each of these two approaches. The oldest is the energy extractive approach. Since we do not have space to review all possible flowmeters, we will only review a few of the more common ones here. Within the energy extractive approach, the differential-pressure-producing flowmeters, such as the orifice plate, the flow nozzle, and the venturi tube, are the most common.

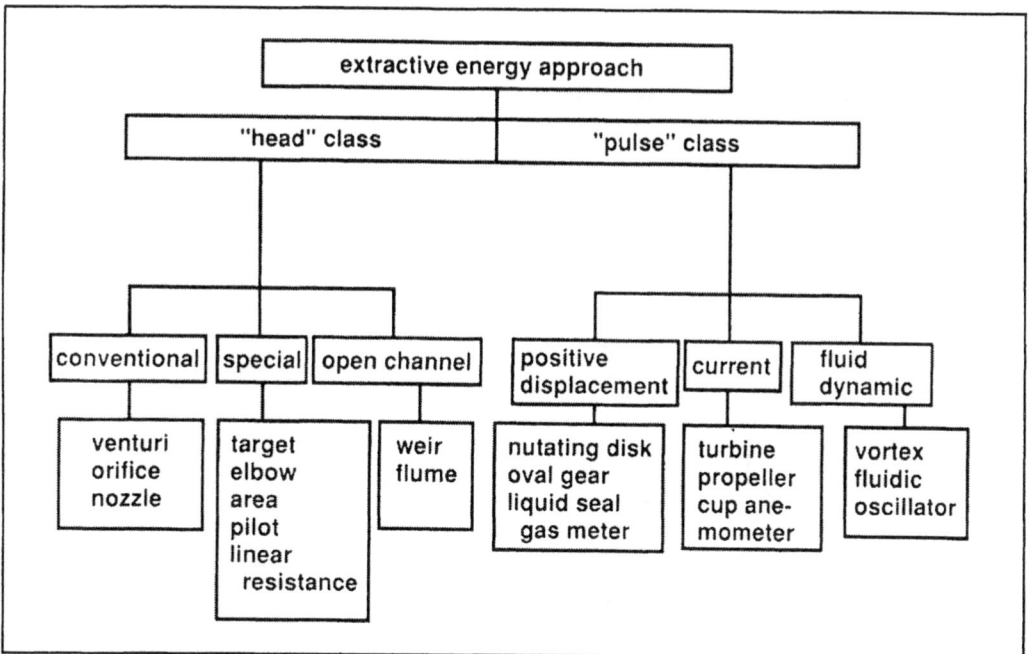

Figure 5-3. Family Tree of Extractive Energy Flowmeters

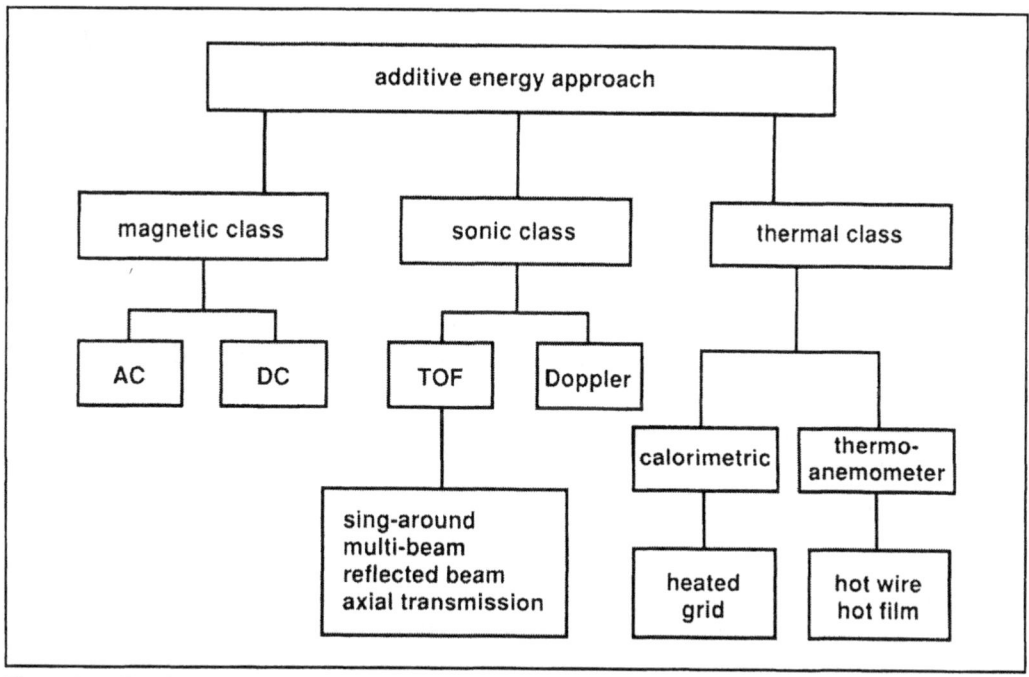

Figure 5-4. Family Tree of Additive Energy Flowmeters

An orifice meter is shown in Fig. 5-5; this is the most common flowmeter in use today. There are two basic elements—the orifice plate and the differential-pressure transmitter. The orifice plate itself is a thin plate with a hole bored in it. It acts to restrict the flow of fluid through the pipe where the meter is installed.

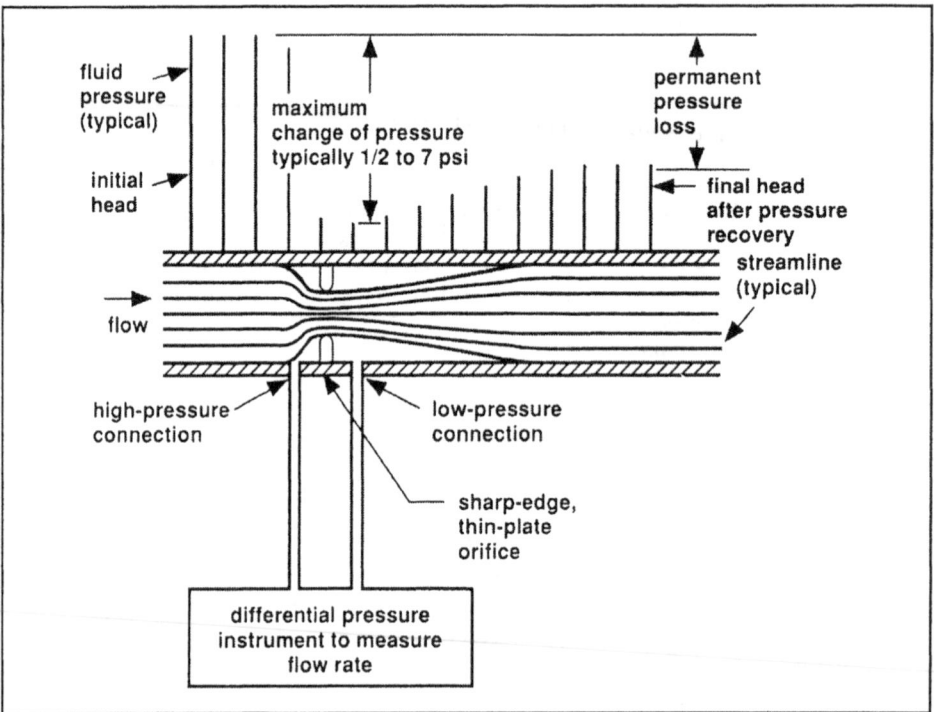

Figure 5-5. An Orifice Meter

As the fluid flows along the pipe and passes through the orifice plate, its velocity must increase because the cross-sectional flow area is decreased. The energy necessary to increase the fluid velocity is acquired by reducing static line pressure. By measuring the change in static line pressure with a differential-pressure transmitter (see the taps for the low pressure connection and the high pressure connection in Fig. 5-5), the volumetric flow rate of the fluid can be inferred. The flow rate is proportional to the square root of the output from the differential-pressure transmitter.

The orifice meter is inexpensive, widely accepted, and readily available. It has a low turndown ratio; it produces a large pressure loss; and its accuracy is affected by plate wear, fluid density (a compressible fluid flow issue), and other factors.

Both the flow nozzle and the venturi tube work on the same principle as the orifice meter, but the restriction shape and flow pattern are different.

5-4.   Other Flowmeters

The magnetic flowmeter uses an energy additive approach. It is based on
Faraday's Law of Magnetic Induction. This law states that a charged
particle passing through a magnetic field produces a voltage that is
perpendicular to both the magnetic field and the velocity vector and that
this voltage is proportional to the velocity of the particle. A magnetic
flowmeter is shown in Fig. 5-6. The liquid, if it is conductive, will contain
charged particles, and its flow through the magnetic field will produce a
voltage that is proportional to the volumetric flow rate. Looking at
Fig. 5-6, the electromotive force of e (or *emf (e)*) is proportional to B, D, and
V. The advantages of the magnetic meter include an obstructionless flow
path and no special pressure loss. Moreover, with proper linings the
flowmeter can be used in corrosive service. Its disadvantages are its cost,
the need for liquid conductivity, and calibration shifts due to electrode
coating while in service.

The vortex flowmeter uses a blunt object in the flow path to generate
vortices. The rate of production of vortices is proportional to the
volumetric flow rate. As these vortices travel through the meter, they
create alternating low- and high-pressure areas. The sensing elements, in
turn, measure these alternating pressures. Typically, an electrical signal is
created at the same frequency as the alternating pressures. This signal is

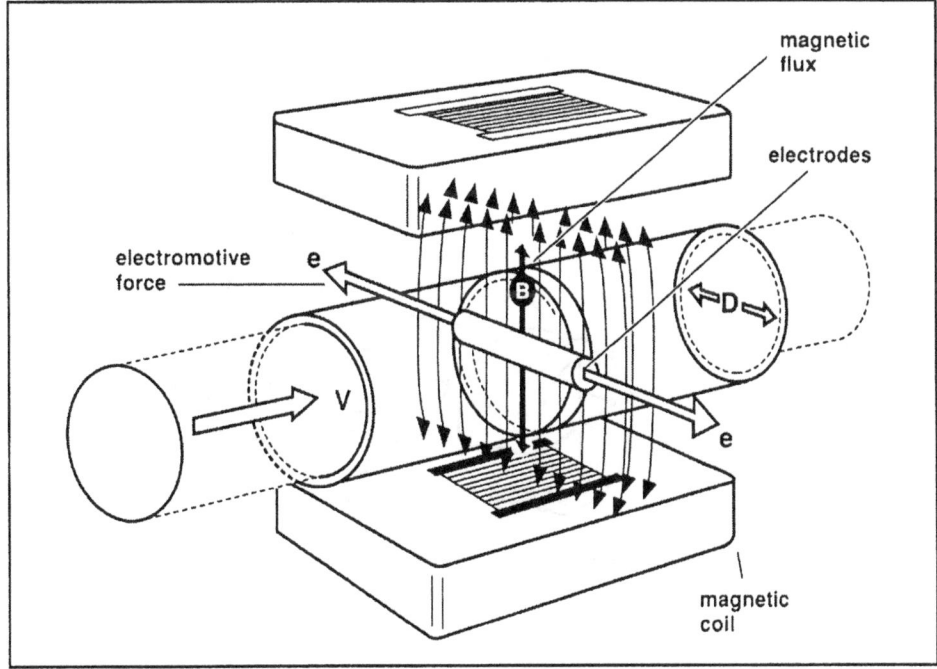

Figure 5-6. A Magnetic Flowmeter

conditioned into a pulse or into an analog signal. A typical vortex meter is shown in Fig. 5-7. These flowmeters are moderate in cost, have a low pressure loss, and are accurate. Their disadvantages are that a blunt body must be based in the flow path, wearing of the object can cause calibration shifts, and some of the alternating pressure sensors are easily damaged by flow overranges.

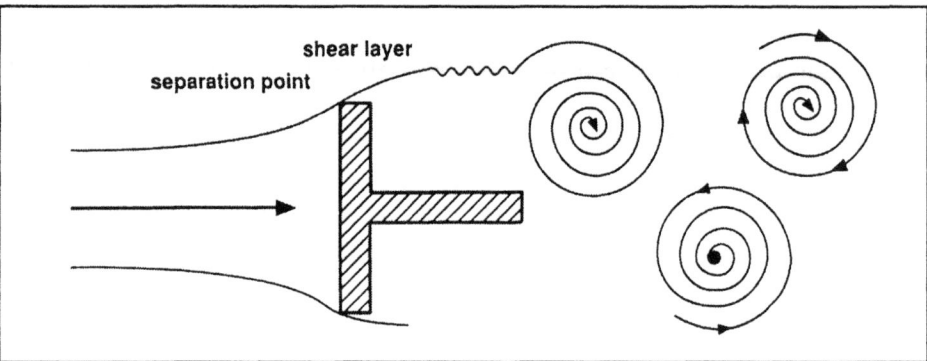

Figure 5-7. A Vortex Flowmeter

The turbine flowmeter, as illustrated in Fig. 5-8, measures volumetric flow by sensing the rotation of a turbine blade in the flow stream. A magnetic pickup is used to measure the rotation and produce a voltage pulse that is at a frequency proportional to the flow rate.

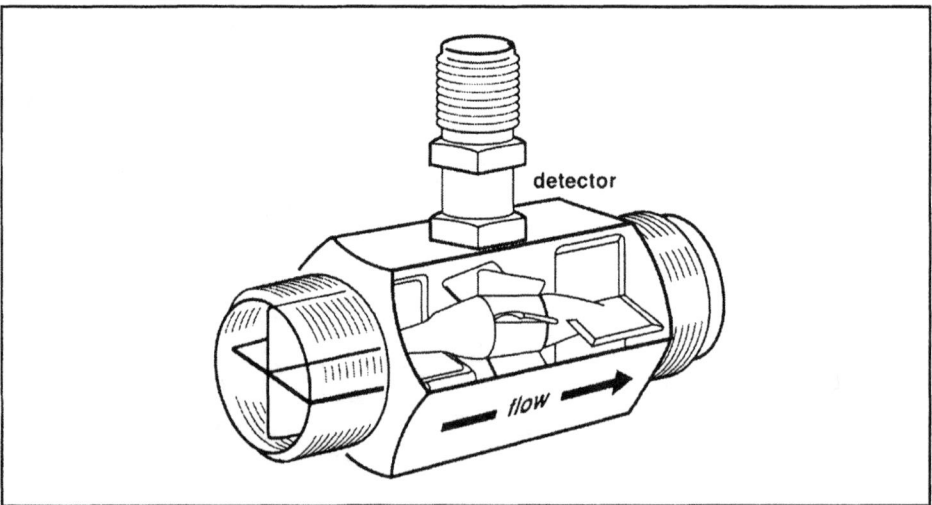

Figure 5-8. A Turbine Meter

Up to this point, we have focused on volume flow rate, Q (typically in gpm):

$$Q = V A \qquad (5\text{-}1)$$

where:

V = velocity of the fluid medium
A = the area of the conduit

Often of more interest than volume flow rate is mass flow rate, M (typically in pounds per minute):

$$M = \rho A V \qquad (5\text{-}2)$$

where ρ = the fluid density.

To determine mass flow rate, we can measure two parameters—volume flow rate and density. This leads to an inferential measurement. An alternative approach is to measure mass flow directly, that is, by a direct mass measurement that is independent of the properties and state of the fluid. Such mass measurements are based upon Newton's Second Law. Some examples are Coriolis-type mass flowmeters, gyroscopic mass flowmeters, and angular-momentum mass flowmeters.

5-5.   Level Sensors/Transmitters

Level sensors can be grouped into seven categories according to the primary level-sensing principle. This is shown in Table 5-1.

(a) Dipstick	
(b) Sight	Glass gage Displacer Tape float
(c) Force	Diaphragm Weighing Bouyancy
(d) Pressure	Hydrostatic head Bubbler Differential pressure
(e) Electric	Capacitance probes Conductance sensors Resistance sensors
(f) Ultrasonic and Sonic Detectors	
(g) Others	Infrared detectors Microwave detectors Nuclear-type detectors Thermal-type detectors

Table 5-1. Level Sensors

The signal produced by one of these measurements must be transduced into an electrical, pneumatic, or digital signal for use in process control. Level sensing is more intuitively understandable than most other process variables, so we will give only a minimum amount of attention to the subject here. A few examples will serve to illustrate the concept.

Fig. 5-9 shows how a differential-pressure transmitter can be used to measure the level of a liquid under pressure. Note that the output of the transmitter depends only upon level and not upon the static pressure in the tank. This is an example of a pressure-type level device.

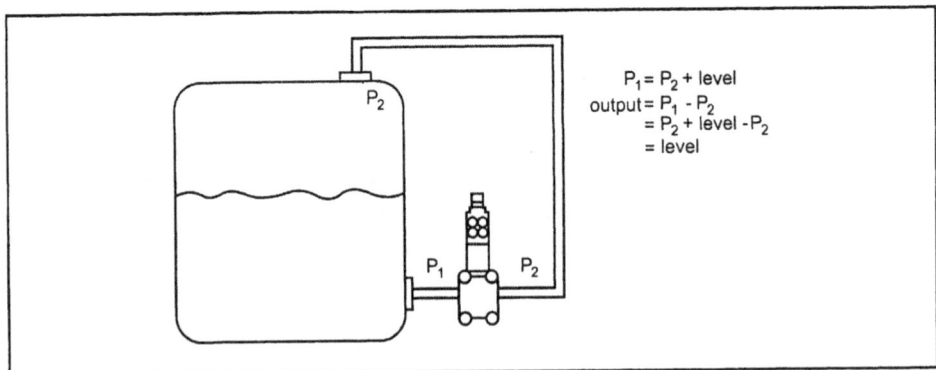

$$P_1 = P_2 + \text{level}$$
$$\text{output} = P_1 - P_2$$
$$= P_2 + \text{level} - P_2$$
$$= \text{level}$$

Figure 5-9. A Differential Pressure Level Measurement

A displacer-type level sensor and its transmitter are shown in Fig. 5-10; this is an example of a force-type level device.

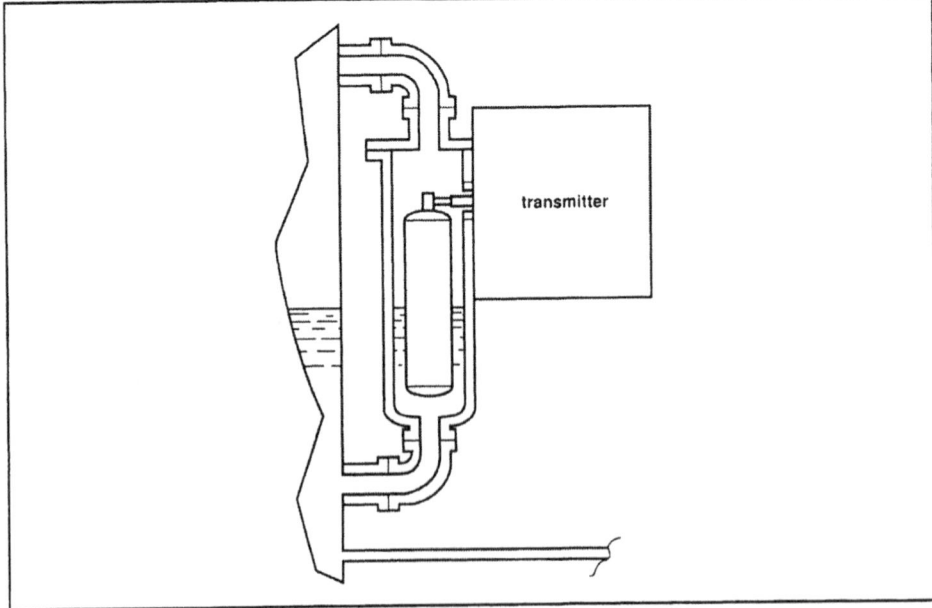

Figure 5-10. A Displacer-Type Level Sensor/Transmitter

Capacitor probes are an example of an electric-type level detector. A probe is inserted into the tank whose level is to be determined. The probe is a variable capacitor; it consists of two conducting materials separated by an insulator. The capacitance of the probe is determined by the area of the conductors, the distance between them, and the dielectric constant of the insulator. Typically, the probe is one electrode, the metallic tank the other. As the level changes, the liquid will cover or uncover the probe, thus changing its capacitance.

5-6.   Temperature Measurement Using Thermocouples

The most commonly used temperature sensors are as follows:

- thermocouples
- resistance thermometers
- filled systems
- radiation pyrometers

The range of temperatures in which these various sensors can be used is shown in Table 5-2.

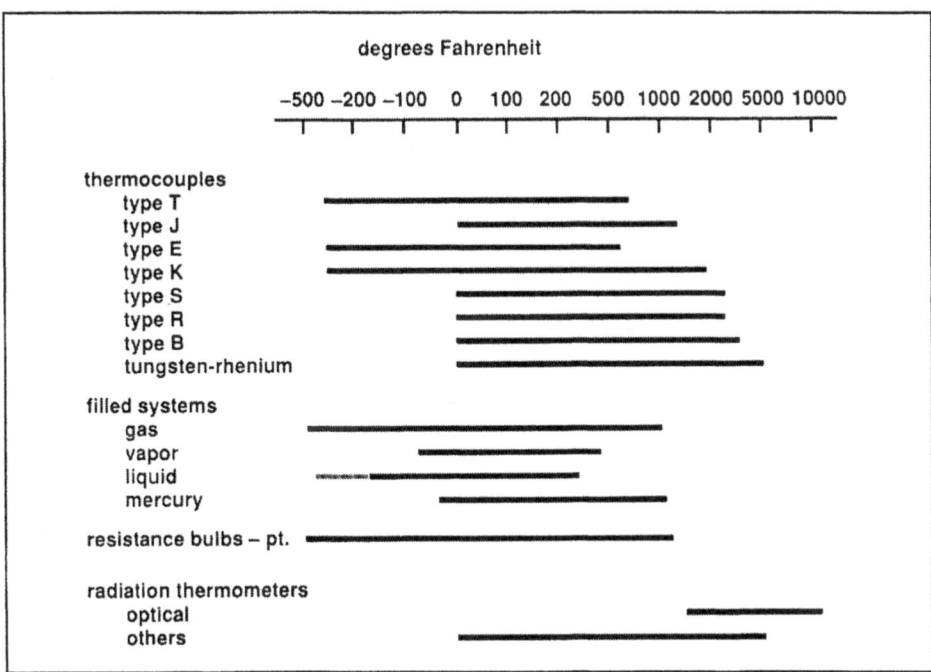

Table 5-2. The Useful Range of Various Temperature Sensors

Easily the most common device for measuring temperature is the thermocouple. It is two wires of unlike metals, joined at one end called the "hot" end or, more properly, the "measuring junction" (see Fig. 5-11). At the cold end, the open circuit voltage, called the *Seebeck voltage*, is measured. This Seebeck voltage depends upon the difference in temperature between the hot and cold junctions and the Seebeck coefficient of the two metals. In Fig. 5-11, the Seebeck current I will flow if terminals a and b are shorted together.

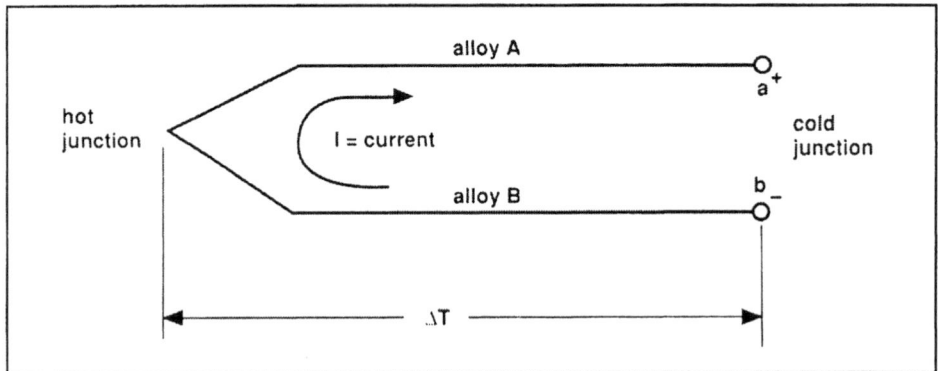

Figure 5-11. Thermocouple Circuit

The Seebeck voltage can be used to indicate temperature only if the cold junction temperature is known or if any changes in cold junction temperature are compensated. In a circuit, this can be done as shown in Fig. 5-12, where a cold junction compensating resistor R_T is provided. Note also that in Fig. 5-12 copper wire is used as "extension" wire for the thermocouple pair.

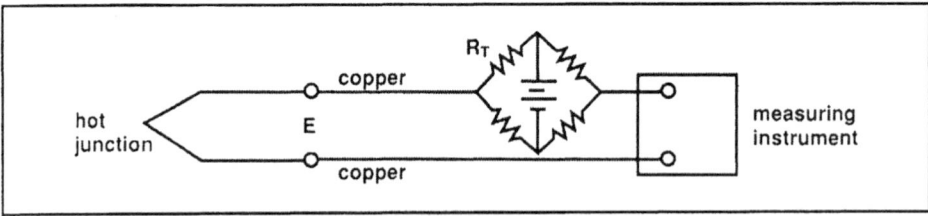

Figure 5-12. Automatic Cold Junction Compensation

The most common thermocouple types are given in Table 5-3. A typical industrial thermocouple assembly is shown in Fig. 5-13.

Thermocouple Type	Temperature Range	Limits of Error	
		Standard*	Premium
T	0–350°C, 0–660°F –200–0°C, –300–32°F	±1°C or ±0.75% ±1°C or ±1.5%	±0.5°C or 0.4% —
J	0–750°C, 0–1400°F	±2.2°C or ±0.75%	±1.1°C or ±0.4%
E	0–900°C, 0–1650°F –200°–0°C, –300°–32°F	±1.7°C or ±0.5% ±1.7°C or ±1%	±1°C or ±0.4% —
K	0–1250°C, 0–2300°F –200°–0°C, –300°–32°F	±2.2°C or ±0.75% ±2.2°C or ±2%	±1.1°C or ±0.4% —
R, S	0–1450°C, 32–2700°F	±1.5°C or ±0.25%	±0.6°C or ±0.1%
B	800–1700°C, 1500–3100°F	±0.5%	—
Tungsten-rhenium 24ga.**	0–2930°C, 32–4200°F	±13.3°C or 1%	—

* Whichever is greater
** Smaller gages have lower temperature limits

Table 5-3. Most Common Thermocouple Types

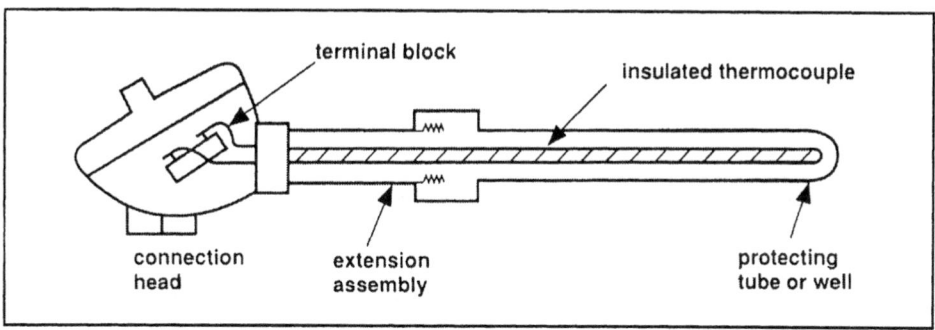

Figure 5-13. A Typical Thermocouple Assembly

5-7. Other Temperature Measurements

Resistance thermometry exploits the relationship between resistance and temperature by using a wire or film with a specific resistance that is relatively high, predictable, and stable. A typical resistance thermometer probe is shown in Fig. 5-14. These probes are provided with two, three, or four leads, depending upon the detection system being used. Three leads is the most common. Wheatstone bridge circuits can be used to measure the resistance and thus the temperature. Modern digital electronic devices operate as shown in Fig. 5-15. In this circuit, a constant current source drives a current through $R_X + 2R_L$. A voltage detector reads a voltage proportional to $R_X + R_L$ at B and a voltage proportional to $R_X + 2R_L$ at A. $V_A - V_B$ is proportional to R_L, so $V_B - V_A + V_B$ is proportional to R_X alone. The measurements are made sequentially, digitized, and stored until the differences can be computed.

Filled-system thermometers have been in use for years. A typical one is shown in Fig. 5-16. A capillary tube connects a bulb containing a fluid that is sensitive to temperature changes to an element that is sensitive to pressure or volume changes. The low cost of electronic systems has caused them to replace filled systems in most process control applications.

Radiation pyrometers measure the energy radiated from an object whose temperature is being measured. Their use in process control is relatively limited.

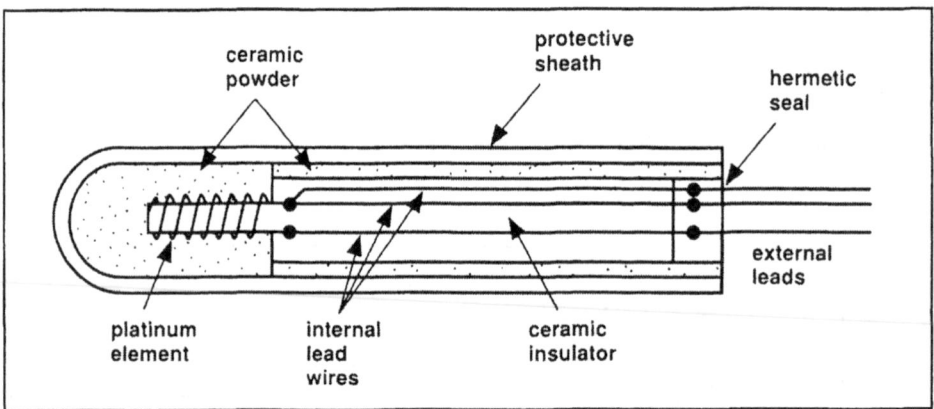

Figure 5-14. A Typical Resistance Thermometer Probe

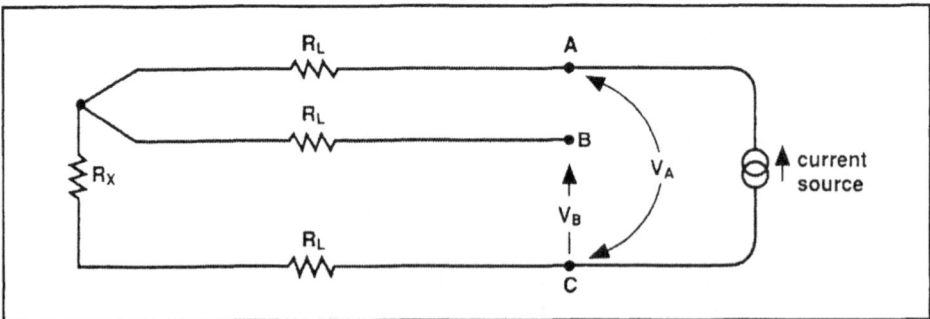

Figure 5-15. Digital Resistance Thermometer Measurement

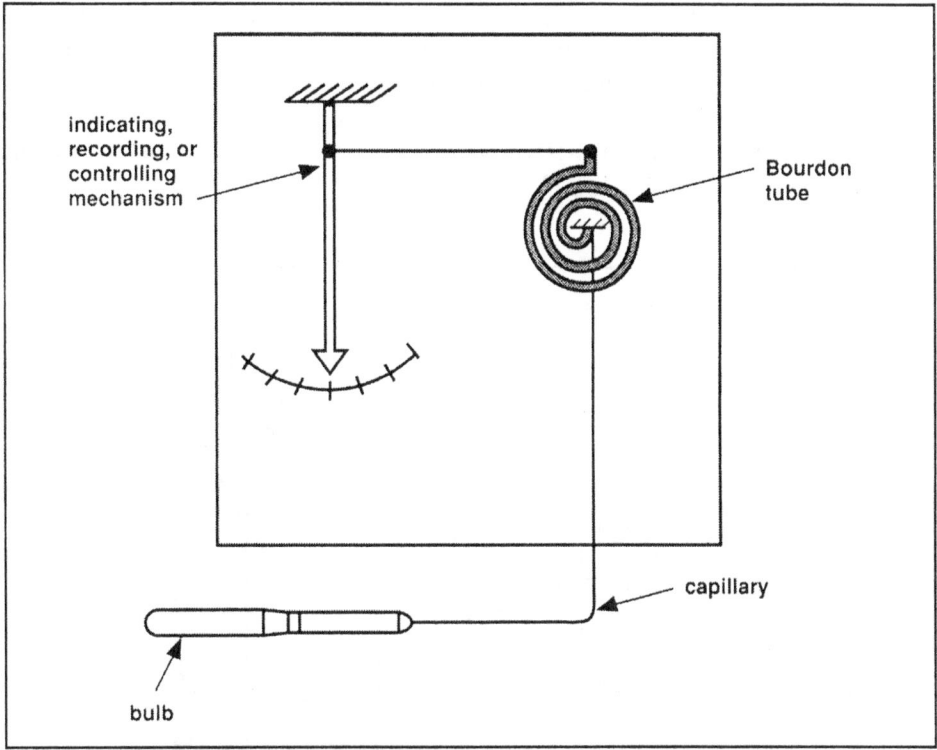

Figure 5-16. A Filled System Thermometer

5-8. Analytical Measurements

Analytical sensors—often called simply *analyzers*—are used to measure composition, usually of a solution of some sort. The focus is on concentration measurement. Most of these measurements are inferential, that is, a measure of some property is used inferentially to measure the concentration of some component in the mixture. There is a wide variety of such sensors, and we will describe a few of them here.

pH Measurement

pH is a measure of whether a solution is acidic or basic; it measures the "potential" of hydrogen ions. Pure water, absolutely neutral, has a pH of 7.0. Less than 7.0 is acidic; greater than 7.0 is basic. The scale runs from 0 to 14. Mathematically, pH is defined as follows:

$$pH = -\log C_H \qquad (5\text{-}3)$$

where C_H = hydrogen ion concentration in gm-moles/liter.

pH is measured by establishing a galvanic cell that develops an electric potential from a chemical reaction at the electrodes.

Two electrodes are used, each having an internal galvanic half cell. One electrode is the measurement electrode, and the other is the reference electrode (see Fig. 5-17). When these two electrodes are immersed in the solution whose pH is to be measured, they can then be externally connected to form a closed circuit. A current will then flow and can be measured to indicate pH.

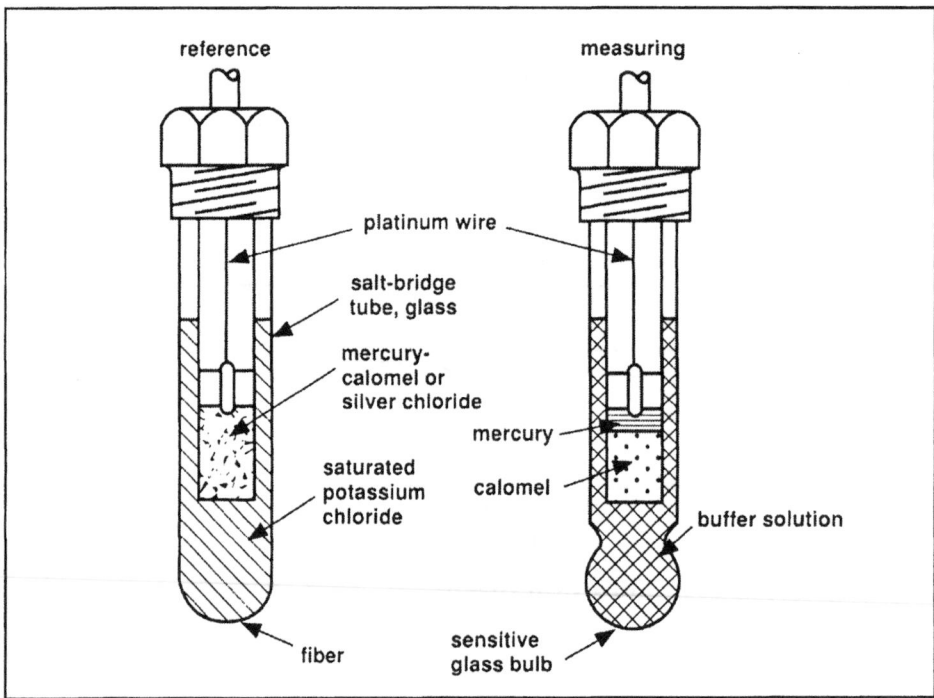

Figure 5-17. pH Measuring and Reference Electrodes

Chromatography

Chromatography is especially useful for measuring the concentration of chemical components in a multicomponent sample. Typically, the sample is vaporized if it is not already in that state. The vapor is then bled into an inert carrier gas stream that flows through a small tube filled with absorbent material. Through repeated absorption and desorption, the sample components will emerge from the tube at different times. A sensor detects each component as it emerges from the tube. In addition to gas chromatography, it is possible to do liquid chromatography.

Infrared

Used to measure concentration, infrared techniques are based upon the fact that different chemical molecules absorb different combinations of frequencies of infrared radiation. The pattern of absorbed frequencies identifies the molecular compound, and the amount absorbed can be used to measure the concentration. Infrared techniques are useful for gases and liquids.

Thermal Conductivity

This technique is useful for gases and vapors. It is based upon the fact that different substances have different abilities to conduct heat. In applying the principle to measurement, the absolute thermal conductivity is not measured. Instead, the conductivity as compared to the conductivity of air is measured at a specific condition.

Others

The list of analyzers is nearly endless, for example, electrical conductivity, oxygen analyzers, flame ionization detectors, carbon dioxide analyzers, mass spectrometers, and so on. The ones discussed here are for illustration purposes only. The serious student certainly needs to explore the subject in much greater depth.

Further Study

ISA publishes and provides numerous books, videotapes, and materials related to measurement. These include videotapes and training courses on measurement, but there are also individual books on flow measurement; level measurement; temperature measurement; pressure, level and density measurement; and so forth. Some examples are the following:

1. R. Annino and R. Villalobos, *Process Gas Chromatography: Fundamentals and Applications* (ISA, 1992).

2. D. Gillum, *Industrial Pressure, Level, and Density Measurement* (ISA, 1995).

3. T. W. Kerlin, *Practical Thermocouple Thermometry* (ISA, 1998).

4. G. K. McMillan, *pH Measurement and Control*, 2d ed. (ISA, 1994).

5. R. E. Sherman, *Analytical Instrumentation*, Practical Guide Series (ISA, 1996).

6. E. Smith, *Principles of Industrial Measurement for Control Applications* (ISA, 1984).

7. D. W. Spitzer, *Flow Measurement*, Practical Guide Series (ISA, 1991).

8. L. M. Thompson, *Electrical Measurements and Calibration: Fundamentals and Applications*, 2d ed. (ISA, 1994).

EXERCISES

5-1.   *A manometer is a device with a transparent U-tube and two liquid surfaces, as shown in the accompanying figure.*

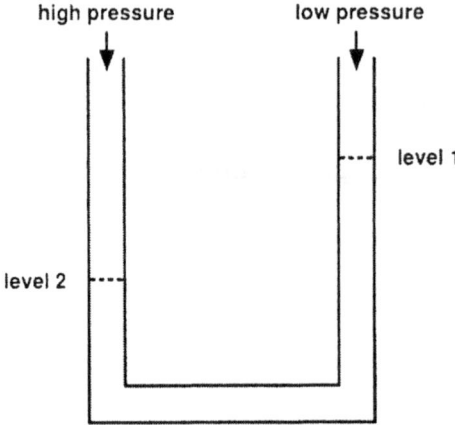

If the low pressure is ambient pressure of 14.6 psi, find the high pressure if the difference in levels is 30 inches of water.

5-2.   *Draw a simple block diagram sketch of a pressure sensor/transmitter.*

5-3.   *Fig. 5-9 illustrated a system to measure level in a closed tank. It is simpler in an open tank. Sketch an open-tank system.*

5-4.   *Archimedes' Principle states that a body fully or partially immersed in a fluid is buoyed up by a force equal to the weight of the fluid displaced. Sketch a simple system to measure level by measuring the weight of a suspended displacer.*

5-5.   *The orifice plate flow equation is as follows:*

$$v = C\sqrt{\frac{\Delta P}{\rho}}$$

or

$$v = C\sqrt{2gz}$$

where:

v = *flow velocity in ft/sec*
C = *orifice flow constant*
g = *gravitational constant = 32.2 ft/sec^2*
ρ = *density, lb/ft^3*
ΔP = *pressure drop, lb/ft^2*
z = *pressure head, ft*

Calculate the fluid velocity for incompressible flow through an orifice plate with a flow constant of 0.6 and with a pressure drop of 16 inches.

5-6. A turbine flowmeter can be used to measure total flow instead of flow rate. Discuss how this can be done.

5-7. The Seebeck voltage is linearly proportional to temperature:

$$Voltage = \alpha \, (T_1 - T_2)$$

where α is the Seebeck coefficient. Calculate the Seebeck voltage for an α of 38 $\mu V/°C$ for junction temperatures of 50°C and 80°C.

5-8. Shown in the following graph is the plot for temperature versus resistance or voltage for a thermocouple, a resistance temperature detector (RTD), and a thermistor (another type of temperature-sensitive resistor).

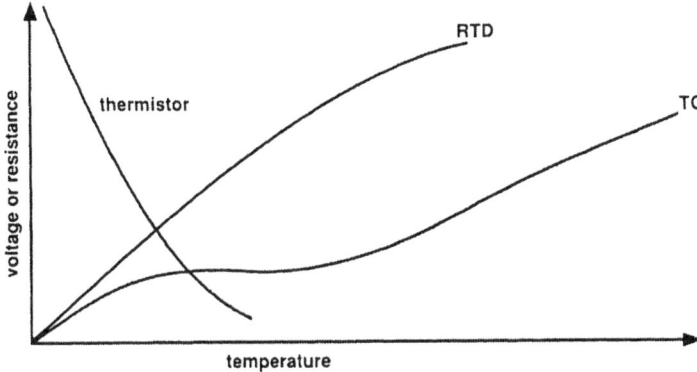

Comment on what these graphs teach us.

Unit 6: Controllers

UNIT 6

Controllers

The feedback controller determines the changes that are needed in the manipulated variable to compensate for disturbances that upset the process or for changes in set point. Understanding the controller's action is a necessity.

Learning Objectives — When you have completed this unit, you should:

A.   be able to explain proportional control action and its advantages and disadvantages

B.   be able to explain reset (integral) action and its advantages and disadvantages

C.   be able to explain rate (derivative) control action and its advantages and disadvantages

6-1.   Controllers

The controller is a special-purpose calculator that uses the error signal from the comparator as its input or forcing function. It calculates the changes needed in the manipulated variable.

Quite often in hardware discussions, the controller case itself and everything inside that case are referred to as the *controller*. But in addition to the controller itself, within this case are many other functional elements of the feedback loop, such as, the input elements, the comparator, the receiver from the feedback transmission system, and, quite often, a recorder. Controllers are often classified according to the chief source of power they use, that is, electronic (including digital), pneumatic, mechanical, or hydraulic. All four of these types have response rates that are rapid enough for conventional process requirements, but in recent years the basic controls used in most process control applications have been either electronic or pneumatic.

The intrinsic safety and simplicity of pneumatic controls are important advantages, but the easy transmission and manipulation of electronic signals gives electronic devices significant advantages. This is especially true when the instrumentation involves significant digital hardware. As a result of these and many other factors, the growth of electronic (including digital) controllers has been the most significant during the past twenty years, and today they command the great bulk of the marketplace.

Rather than concentrate extensively on the detailed design of specific pieces of hardware, however, the general purpose of this unit will be to help you understand the various modes of control action.

6-2.   On-Off Control

Every type of control action may be considered as either continuous or discontinuous control. In one sense, any digital control is a special case of discontinuous control, but when practitioners refer to discontinuous control action they usually are speaking of "two-position" or "multiposition" control. Our primary focus in this section will be on two-position control.

Two-position control action, or *on-off control*, is one of the most widely used types of control for both industrial and domestic service. Many of us are familiar with this type of control action since it is used on most home heating systems and most domestic hot-water heaters.

Two-position control is a type of control action where the manipulated variable is quickly changed to either a maximum or a minimum value, depending on whether the controlled variable is greater or less than the set point or some bandwidth around the set point. The minimum value of the manipulated variable is usually zero ("off").

The mechanism for generating on-off control is usually a simple relay. In conventional practice, it is not possible to build a device that is sensitive to the sign of extremely small deviations, and--more importantly--it is not desirable to do so. Such an excessively sensitive controller would undergo needless wear and tear on its moving parts and contacts and would keep the process very unsteady. The solution in most commercial two-position controllers is to establish a *dead zone* of about 0.5 percent to 2.0 percent of full range. The terms *differential gap* and *neutral zone* are often used synonymously with *dead zone*. This dead zone, of course, straddles the set point; no control action takes place when the control variable lies within the dead zone itself. Fig. 6-1 shows what two-position (without a dead zone) control for a home heating system might look like.

The instruments used for on-off control are cheap, rugged, and virtually foolproof. On-off control is inherently oscillatory in character, but for many systems the amplitude of such oscillation of the controlled variable can be quite small.

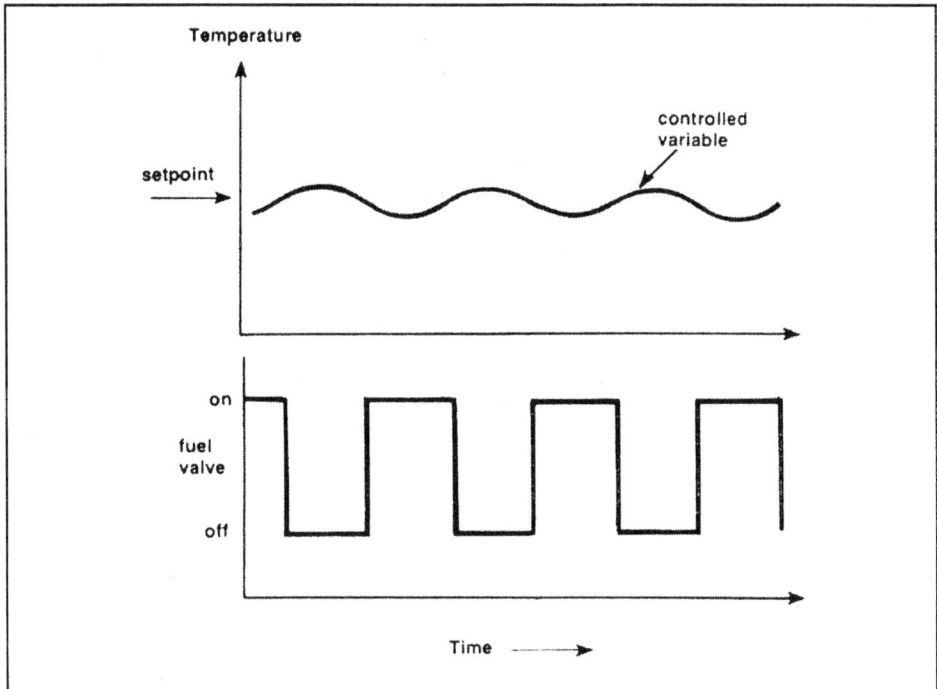

Figure 6-1. Two-Position Control of a Heating System

A variant of two-position control with differential gap is three-position control, wherein the controller responds with an intermediate output when the controlled variable lies within the neutral zone. Fig. 6-2 illustrates this kind of control and its response characteristics.

Commercial controllers with additional steps are available. Commercial controllers with as many as three intermediate output positions (this is effectively five-position control) are available, but such multiposition control is not used extensively.

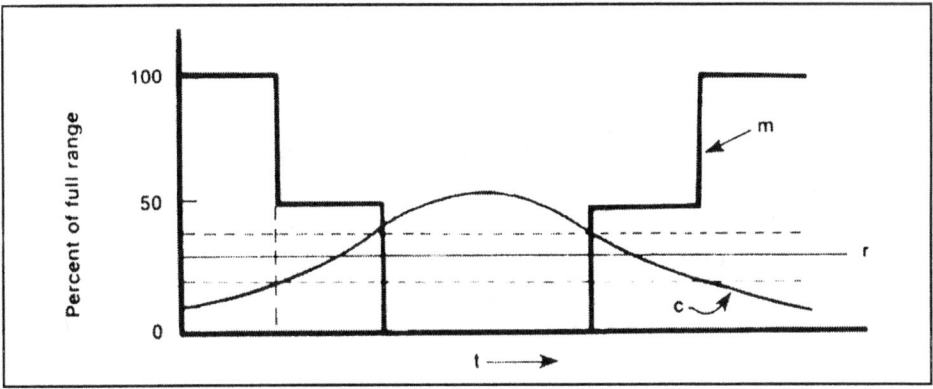

Figure 6-2. Three-Position Control

6-3.   Proportional Control

The most basic continuous control mode is *proportional control* where the controller output is algebraically proportional to the error input signal to the controller. This is illustrated in the following simple block diagram model of the controller, which is subjected to an input error step change of height *E*:

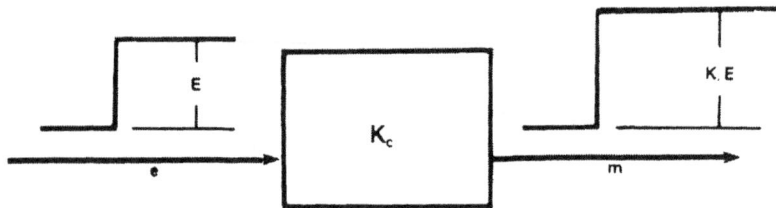

In this case, the controller output *m* is calculated as follows:

$$m = K_c e \tag{6-1}$$

This equation is called the control *algorithm*.

Proportional control is the simplest and most commonly encountered of all the continuous control modes. In effect, there is a continuous linear relationship between the controller input and output. There are several additional names for proportional control, such as *correspondence control* (because of the linear correspondence of output to input), *droop control* (because of the droop or offset characteristic, which will be discussed shortly), and *modulating control* (because of the proportional adjustment).

The *gain* of the controller is the term K_c. This is also referred to as the *proportional sensitivity* of the controller. It indicates the change in the manipulated variable per unit change in the error signal. In a very true sense, the proportional sensitivity or gain is an amplification and represents a parameter on a piece of actual hardware that must be adjusted by the operator. In effect, the gain is a knob to be adjusted (or a number to change in a line of software code).

The gain-adjusting mechanism on many industrial controllers is not expressed in terms of proportional sensitivity or gain but in terms of *proportional band* (PB). Proportional band is defined as the span of values of the input that corresponds to a full or complete change in the output. This is usually expressed as a percentage and is synonymous with *throttling band* or *throttling range*. It is related to proportional gain by the following:

$$PB = (1/K_c) * 100 \tag{6-2}$$

Since most controllers have a scale that indicates the value of the final controlled variable, the proportional band can be conveniently expressed as the range of values of the controlled variable that corresponds to the full operating range of the final control valve. This full operating range of the final control valve is often inferred to be the operation of a final control valve through a full stroke.

As a matter of practice, wide bands (high percentages of PB) correspond to less sensitive response, and narrow bands (low percentages of PB) correspond to more sensitive response. Several graphic means are used to illustrate the effects of varying proportional bands, and examples of these means are shown in Figs. 6-3 and 6-4.

Proportional control is quite simple and the easiest of the continuous controllers to tune, that is, there is only one parameter to adjust. It also provides good stability, very rapid response, and is relatively stable dynamically. Proportional control has one major disadvantage, however. At steady state, it exhibits *offset*, that is, there is a difference at steady state between the desired value or set point and the actual value of the controlled variable. This is illustrated in Fig. 6-5.

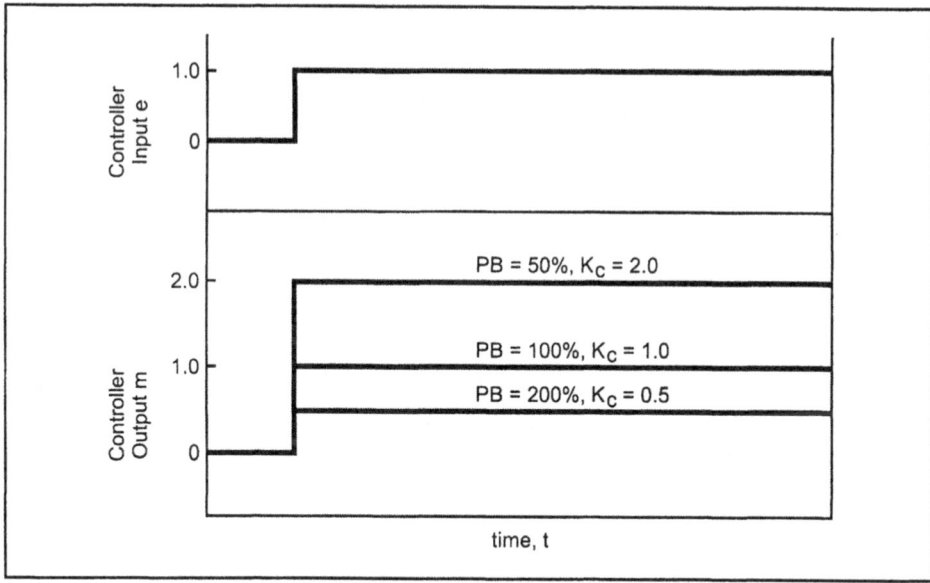

Figure 6-3. Effect of Proportional Control on Controller Output

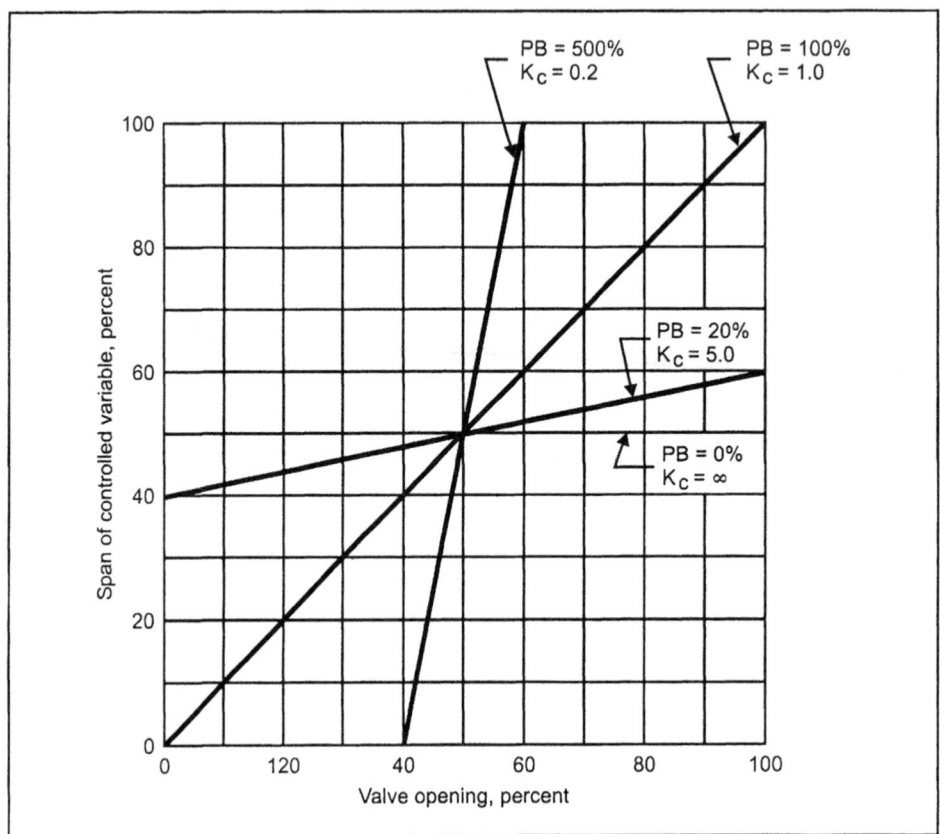

Figure 6-4. Effect of Proportional Band on Valve Opening

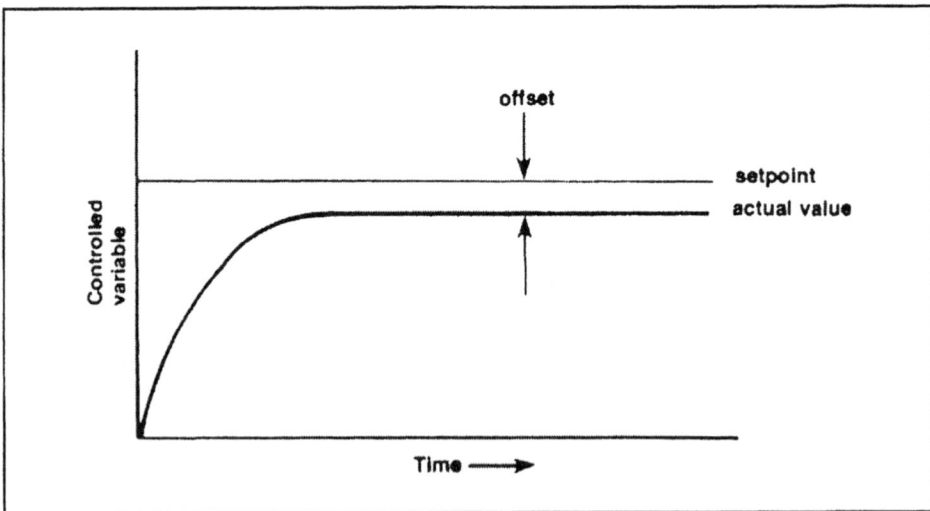

Figure 6-5. Proportional Offset as Seen on a Feedback Recorder

6-4.   Reset Control

Reset control is really an integration of the input error signal *e*. In effect, this means that in reset control (often called *integral control*), the value of the manipulated variable *m* is changed at a *rate* that is proportional to the error *e*. Thus, if the deviation is doubled over a previous value, the final control element is moved twice as fast. When the controlled variable is at the set point (no deviation), the final control element remains stationary. In effect, this means that at steady state, when reset control is present there can be no offset. The steady-state error must be zero.

The use of the word *rate* in explaining reset control creates its own problems. Perhaps it is helpful to think in terms of the historical accumulation of error, with positive errors running up the count with time and negative errors reducing the accumulation. The net accumulation at any time becomes the reset contribution to the manipulated variable.

Reset or integral control action usually is combined with proportional control. The combination is termed *proportional-reset* or *proportional-integral control*. It is usually referred to as *PI control*. The combination is favorable in that some of the advantages of both types of control are available. The basic control algorithm for proportional-plus-integral control is as follows:

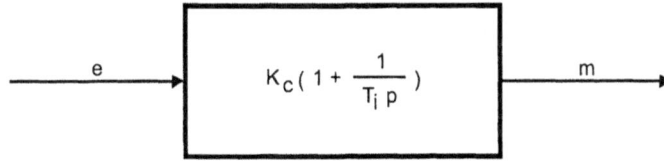

where:

$$K_c = \text{the proportional gain}$$
$$T_i = \text{the integral time}$$
$$e = \text{the error signal}$$
$$m = \text{controller output}$$

p implies the action of taking the derivative with respect to time, *d/dt*; therefore 1/*p* implies integration with respect to time $\int \dots dt$. *

Although this type of block diagram representation for PI control has its advantages, it is desirable to expand it further. Fig. 6-6 gives a detailed breakdown of the way a PI controller might function. In this particular

---

* These terms, of course, are taken from standard calculus, and for the reader who has not been introduced to such mathematics, some difficulties are inherent. With all apologies, however, at this point the mathematics cannot be avoided. *p* is the Heaviside operator *d/dt* and *1/p* implies $\int \dots dt$.

case, the controller block diagram is separated into two parts, the upper part to show the proportional action and the lower part to illustrate the integral or reset action. A step input is provided to the controller, and each of the control modes has its own characteristic output.

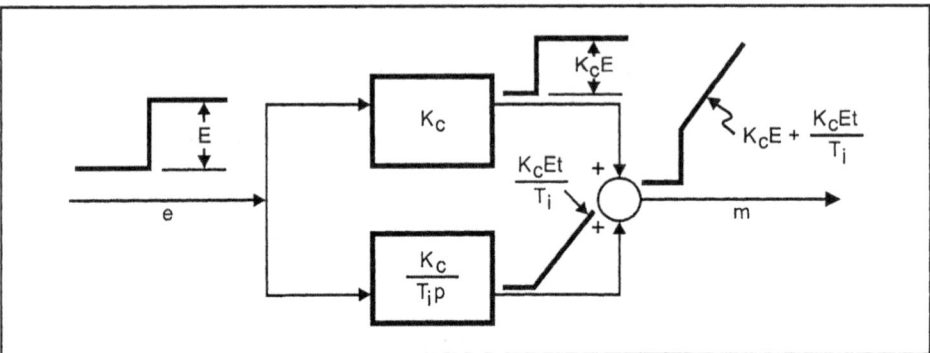

Figure 6-6. Proportional-plus-Reset Controller's Response to a Step Change in Error

The total output of the controller is, of course, the sum of the output of the two individual modes. This is also illustrated in Fig. 6-6. Note that at the end of T_i units of time, the reset mode has tended to repeat (in magnitude of output) the proportional mode. On some controllers, the adjustable parameter for the reset mode is T_i, the *reset time*, and on others it is the reciprocal of T_i, which is referred to as *repeats per minute*. Repeats per minute is also referred to as the *reset rate*.

The advantage of including the integral mode with the proportional mode is that the integral mode eliminates offset. Typically, there is some decreased stability due to the presence of the integral mode; that is, the addition of the integral mode makes the total loop slightly less stable.

One special case is in liquid-flow control loops, which are extremely fast and quite often tend to be very noisy. There are two ways to deal with this problem. One approach is to use a high-proportional band (low gain) to make the loop more stable (sluggish) in the presence of a noisy input signal. A second approach is to add integral control to the feedback controller, even when you do not care about the presence of offset, and count on integral control to provide a dampening or filtering mode for the loop. Of course, the advantage of eliminating any offset is still present, but it may not be the principal motivating factor for the addition of reset.

Tuning a PI controller is more difficult than tuning a simple proportional controller because now there are two separate tuning adjustments that must be made, and each depends on the other. As a matter of fact, the difficulty posed by tuning a controller increases dramatically with the number of adjustments that must be made. The subject of tuning industrial controllers is covered extensively in Unit 9.

It is possible to use integral or reset control by itself without proportional control. This is not a common situation, however, so we will not discuss it further.

6-5.   Rate Control

It is conceivable to have control that is based solely on the rate of change of the error signal e.   However, though this is theoretically possible, it is not practical because, while the error might be huge, if it were unchanging the controller output would be zero. Thus, *rate control* (often called *derivative control*) is usually found in combination with proportional control. The following is a typical descriptive block diagram for a proportional-rate controller or PD controller:

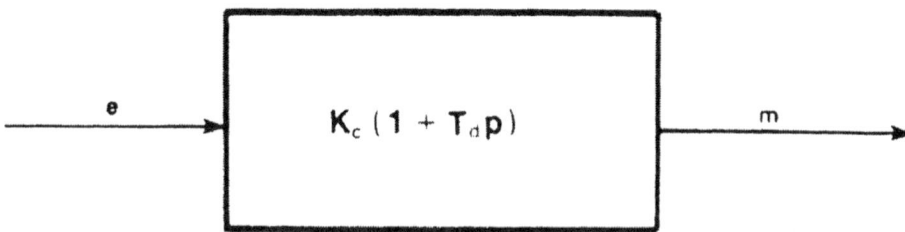

$$K_c \, (1 + T_d p)$$

All of the terms are as defined earlier. In addition, T_d is the *derivative time* and, as before, p implies the operation of taking the derivative with respect to time.

By adding derivative mode to the controller, *lead* is added in the controller to compensate for *lag* around the loop. Almost any process has lag around the loop, and, therefore, the theoretical advantages of lead in the controller are appealing. It is quite a difficult control mode to implement and adjust, however, and its usage is limited to cases where there is an extensive amount of lag in the process. This often occurs with temperature sensors in large vessels. One concern to keep in mind when implementing derivative control is that signal noise tends to drive the derivative mode output inappropriately. Some form of signal filtering is needed within the hardware or software implementation. Many modern controllers base their derivative control on the input signal and not the error signal.

To gain some insight into derivative mode, refer to Fig. 6-7. The block diagram of the controller is broken into two parts to illustrate the separate action of the derivative mode and the proportional mode.

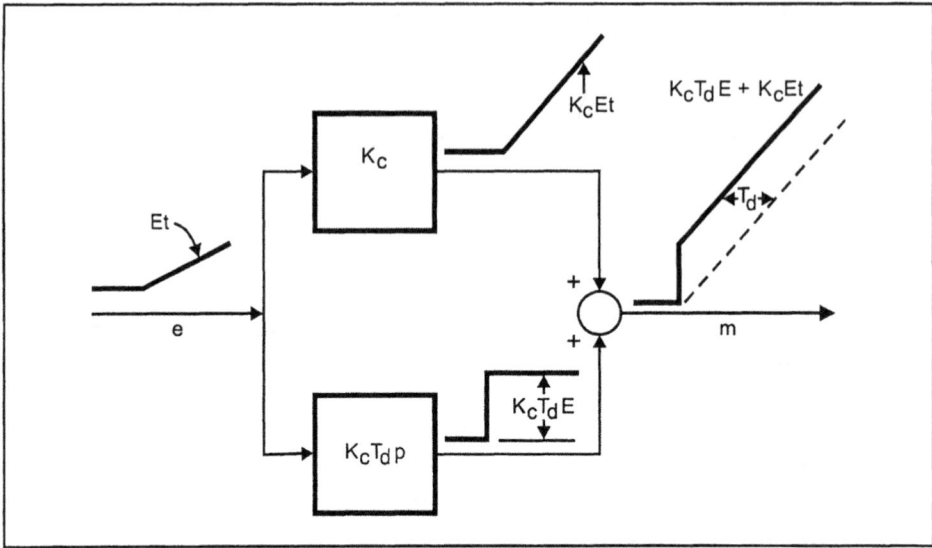

Figure 6-7. The Output of a Proportional-plus-Derivative Controller for a Ramp Input

In Fig. 6-7, the input to the controller is shown to be a ramp change in error signal. By inspecting the combined output of both control modes, it is possible to see that the derivative time T_d, which is adjusted in the controller, is really an adjustment of the amount of lead that is introduced in the controller. The addition of rate control to the controller makes the loop more stable if it is appropriately tuned. Since the loop is more stable, the proportional gain may be higher, and thus it can decrease offset more than can proportional action alone (but, of course, it does not eliminate offset).

6-6.   PID Control

Proportional-plus-integral-plus-derivative control, or *three-mode* control, is the most sophisticated, continuous controller available in feedback loops. Fig. 6-8 shows a typical block diagram for such a PID controller:

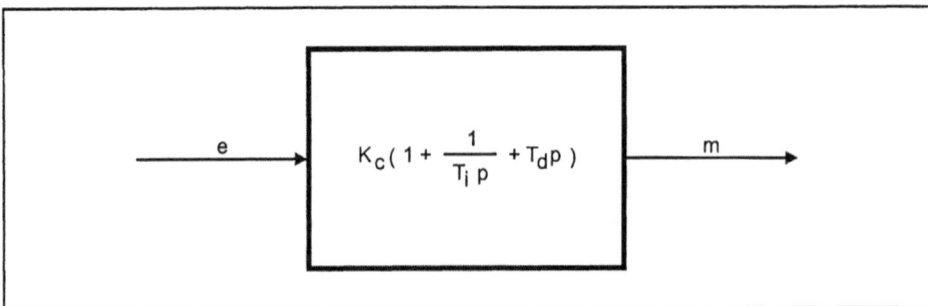

Figure 6-8. Block Diagram for a PID Controller

In three-mode control, we have the most complex controller algorithm that is available routinely. It gives rapid response and exhibits no offset, but it is very difficult to tune—now there are three knobs to adjust. As a result, it is used only in a very small number of applications, and it often requires extensive and continuing adjustment to stay properly tuned. It does offer very good control when good tuning is implemented.

6-7. Summary

By way of review, we will summarize the different conventional control modes presented in this unit. Table 6-1 gives the basic descriptions of the various conventional controller modes currently available. Table 6-2 illustrates the response action of these various control modes to different types of input-forcing functions, that is, to different changes in the input error signal. Table 6-3 provides a summary of the basic characteristics of these controllers.

Symbol	Description	Mathematic Expression
		One Mode
P	Proportional	$m = K_c e$
I	Integral (reset)	$m = \dfrac{1}{T_i} \int e\, dt$
		Two Mode
PI	Proportional-plus-integral	$m = K_c \left[e + \dfrac{1}{T_i} \int e\, dt \right]$
PD	Proportional-plus-derivative	$m = K_c \left[e + T_d \dfrac{d}{dt} e \right]$
		Three Mode
PID	Proportional-plus-integral-plus-derivative	$m = K_c \left[e + \dfrac{1}{T_i} \int e\, dt + T_d \dfrac{d}{dt} e \right]$

Table 6-1. Conventional Controller Modes

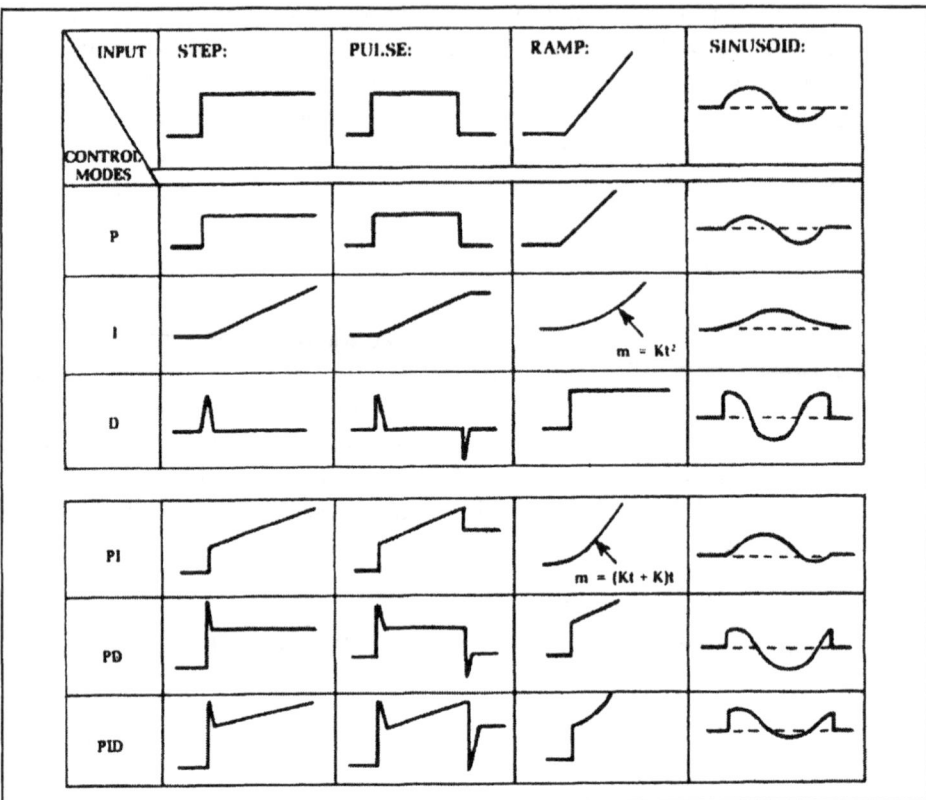

Table 6-2. Response of Controller Modes

Two-Position:
    Inexpensive
    extremely simple

Proportional:
    Simple
    Inherently stable when properly tuned
    Easy to tune
    Experiences offset at steady state

Proportional-plus-reset:
    No offset
    Better dynamic response than reset alone
    Possibilities exist for instability due to lag introduced

Proportional-plus-rate:
    Stable
    Less offset than proportional alone (use of higher K_c possible)
    Reduces lags, i.e., more rapid response

Proportional-plus-reset-plus-rate:
    Most complex
    Most expensive
    Rapid response
    No offset
    Difficult to tune
    Best control if properly tuned

Table 6-3. Characteristics of Controller Modes

EXERCISES

6-1. *For a proportional controller, answer the following questions:*

    a.   *What gain corresponds to a proportional band of 200 percent?*

    b.   *What proportional band corresponds to a gain of 0.2?*

6-2. *Proportional-only controllers exhibit offset at steady state. How is this possible? (Offset implies an error; therefore, why doesn't the controller output eliminate the error?)*

6-3. *In this unit, the error is shown as the input to the controller. Does it make any difference if the error is caused by a change in set point in one case or by a disturbance in another?*

6-4. *Proportional control is often described by the equation $m = K_c e + M_r$ where M_r is an adjustable "manual reset" that can be used to bias the output so as to eliminate offset at a specific operating point. Will this eliminate offset at other operating points? Explain your answer.*

6-5. *It has been proposed that a controller might be designed to check the magnitude of the error. If the error is large the controller would operate as a proportional-only controller; if the error is small, it would operate as a reset-only controller. What would be the advantage(s) of this type of controller?*

6-6. *The "ideal" PID controller was described in Fig. 6-8:*

$$m = K_c(1 + 1/T_i p + T_d p)e$$

But some have proposed that it would be better if it were described as follows:

$$m = K_P e + (K_I/p)e + K_D pe$$

where:

K_P = *proportional gain, equal to K_c*
K_I = *integral gain, equal to K_C/T_i*
K_D = *derivative gain, equal to $K_C T_d$*

The latter form is referred to as "non-interacting." What operating advantages, if any, would it possess?

6-7. *In practice, all controllers include some filtering of the error signal before or within the rate mode. Why is this necessary?*

6-8. *The following block diagram illustrates a PID controller whose input error changes in a stepwise manner. Draw the shape and give the equation of the output for each mode and for the total controller output:*

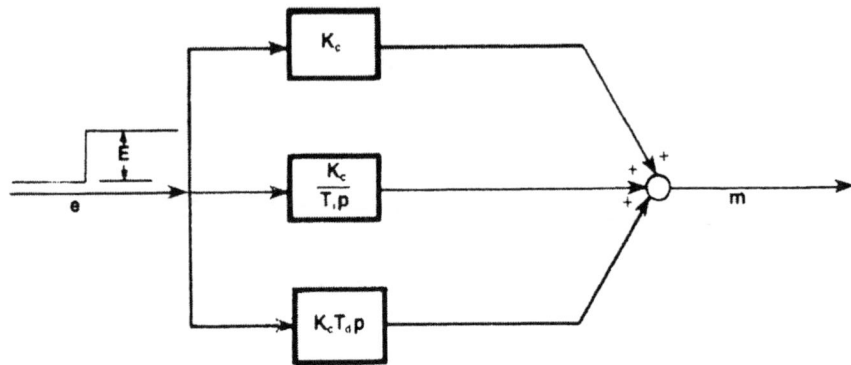

6-9. *Repeat Exercise 6-8 for a ramp input of the following form:*

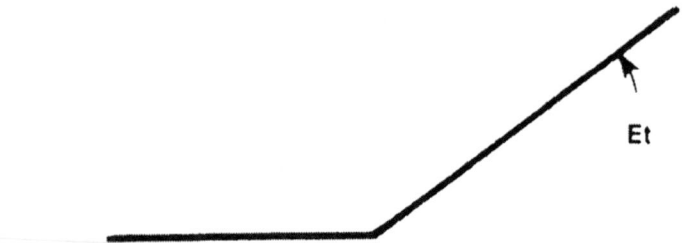

Unit 7: Control Valves

UNIT 7

Control Valves

The output of the controller is a signal to the final control element that adjusts the manipulated variable. In the vast majority of process control applications, the final control element is a valve. This most important piece of hardware merits understanding.

Learning Objectives — When you have completed this unit, you should:

A. understand the purpose and use of control valves, valve actuators, and valve positioners

B. be able to define rangeability and turndown ratio

C. know the meaning and use of valve coefficients and appreciate sizing considerations

D. understand the factors that influence the dynamic behavior of control valves

7-1. Control Valves, Actuators, and Positioners

For most process control systems, the final control element is a valve. These valves are usually driven by motors that are commonly called *actuators*. Actuators are classified on the basis of their power source, and valves are classified on the basis of their body style and flow characteristics. There is almost an unlimited number of hardware variations in process control valves and actuators, and no attempt will be made in this unit to give an exhaustive presentation of the hardware itself. There are several books specifically on control valves, and the student is referred to them at the end of the unit for "Further Study." It will be helpful, however, if we undertake a limited study of control valves here. An example of a control valve is given in Fig. 7-1. This is a sliding-stem, single-seat control valve body with a pneumatic actuator.

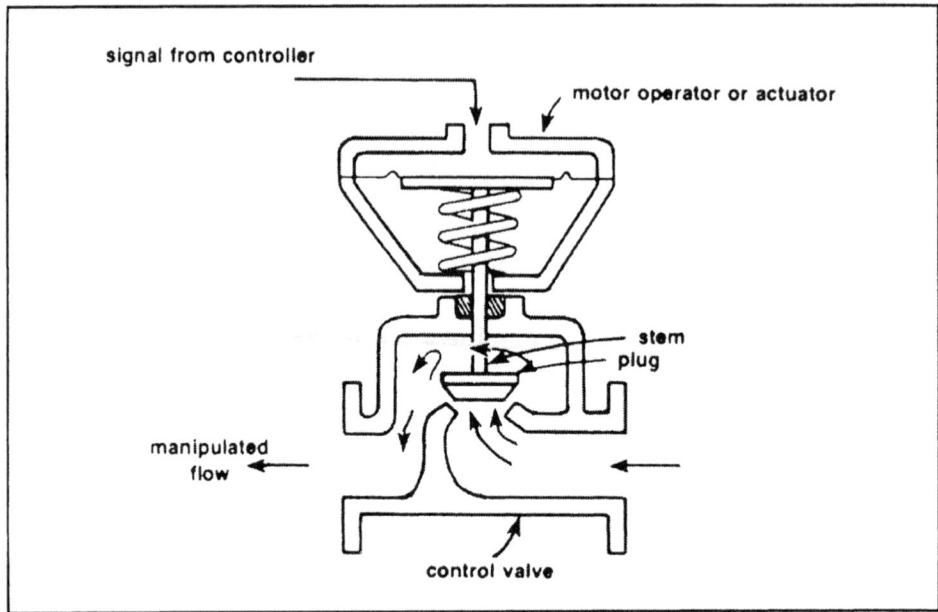

Figure 7-1. A Typical Sliding Stem Control Valve with a Pneumatic Actuator

It is convenient to classify control valve bodies as one of the following:

    a.   linear-stem motion control valves

    b.   rotary control valves

    c.   ball control valves (actually a form of rotary)

We will discuss the most popular members of each of these categories in this unit.

The control valve actuator is used to translate the specific output signal from the controller into a more powerful form. This typically means that it drives the stem of the control valve. Control valve actuators may be pneumatic, electric, hydraulic, or manual. The most commonly used valve motors or actuators in the process industries are pneumatic, and most of these have diaphragm motors. The diaphragm is spring loaded in opposition to the driving air pressure so that the valve stem position is proportional to air pressure. Usually, the diaphragm is made of a rubber-based fabric or other limp material and is supported by a backup plate. With diaphragm motors, the maximum available *stroke*, or stem travel, is usually 2 to 3 inches. For longer strokes, the valve actuator may be a double-acting piston, or it may be a rotary pneumatic device. Electric actuators are not as popular because of their cost and complexity; electrohydraulic actuators are used primarily in areas where no operating air is available.

A *valve positioner* is actually a control valve accessory that transmits a loading pressure to an actuator in order to position a valve plug stem exactly as dictated by the signal from the controller. The positioner is a controller in itself, proportional only. It is often desirable to use a valve positioner to improve both the dynamic and the static behavior of the valve.

A valve positioner is typically an "air relay" that is used between the controller output and the valve diaphragm. It usually has a separate air supply and a feedback signal that indicates the stem position. The positioner acts to eliminate hysteresis, packing-box friction, and valve plug unbalance due to pressure drop of the manipulated flow. It also assures that the valve stem is positioned exactly in accordance with the controller output. The valve positioner is also useful for minimizing the lag associated with the valve response.

7-2.   Linear-Stem Motion Control Valves

Linear-stem motion control valves resemble the globe valve shown in Fig. 7-1 in that the valve plug is positioned by the stem that slides through a packing gland. Typically, linear-stem motion control valves may be single-seated globe, double-seated globe, or gate valves. Each of the various types of valve plugs has advantages and disadvantages, and you will find the books devoted strictly to control valves are a good source for understanding these various characteristics, advantages, and disadvantages.

The flow characteristics of linear-stem motion control valves generally fall into three broad categories based on the relationship between the amount of valve position, travel, stroke, or lift in the opening direction compared to the resulting flow. These categories, illustrated in Fig. 7-2, are as follows:

a.   *Decreasing-sensitivity* type. In this case, the valve sensitivity (defined as the change in flow for a given change in valve position) decreases with increasing flow. (This gives a quick-opening effect.)

b.   *Linear* type. In this case the valve sensitivity is more or less constant throughout the flow range.

c.   *Increasing-sensitivity* type. The most common example is the *equal-percentage* type valve. It derives its name from the fact that the valve sensitivity at any given flow rate is a constant percentage of that given flow rate. Other common terms used are logarithmic or parabolic type valve.

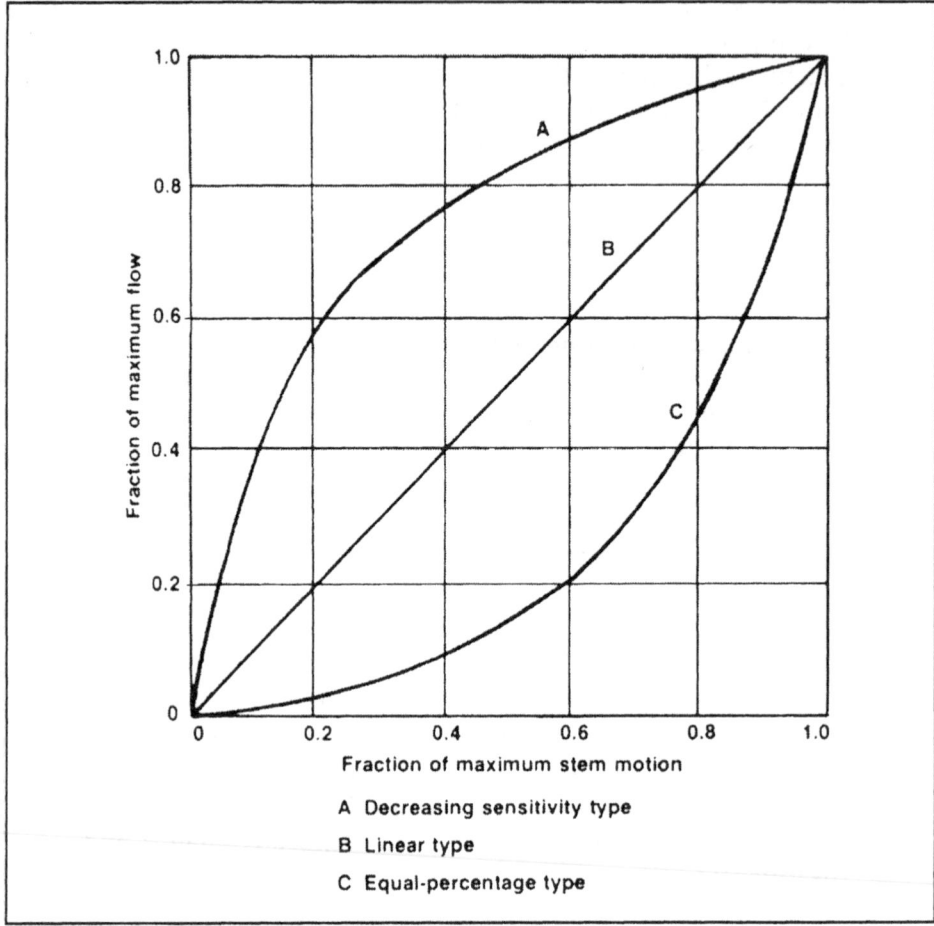

Figure 7-2. Valve Flow Characteristics

Of course, the individual flow characteristics of the wide range of hardware that is available will not always fall into such neat categories. These three types will, at least, provide the practitioner with a systematic way to approach the literally hundreds of valve trim arrangements that are available.

Linear-stem motion control valves have many different body styles. The most common takes the form of a globe. These valves may be either single-seated or double-seated; examples are shown in Fig. 7-3. Single-seated valves are commonly employed in situations where tight shutoff is required or for valves with sizes of 1 inch or smaller.

Double-seated valves generally experience leakage through the valve that is somewhat greater than in single-seated valves. This is because it is virtually impossible to close the two ports simultaneously, especially when thermal expansion and other factors are considered. The advantage of double-seated bodies, however, is that the hydrostatic effect of the fluid pressure acting on each of the two seats will tend to cancel each other out, and much less actuator force is necessary.

Linear-stem motion valves are also used in three-way valve bodies where the control valve may be used to divert a stream or to combine streams. Linear-stem motion valves are also commonly encountered in angle valve situations. In such cases, the valve body is typically single-seated. In addition, linear-stem travel valves are built in Y-style bodies, in split-body styles, and in cage styles. Cage valves usually are designed so the valve trim can be easily removed to facilitate maintenance and replacement, if necessary. Linear-stem motion valves are also used in sliding-gate valve applications and in a number of similar, special applications.

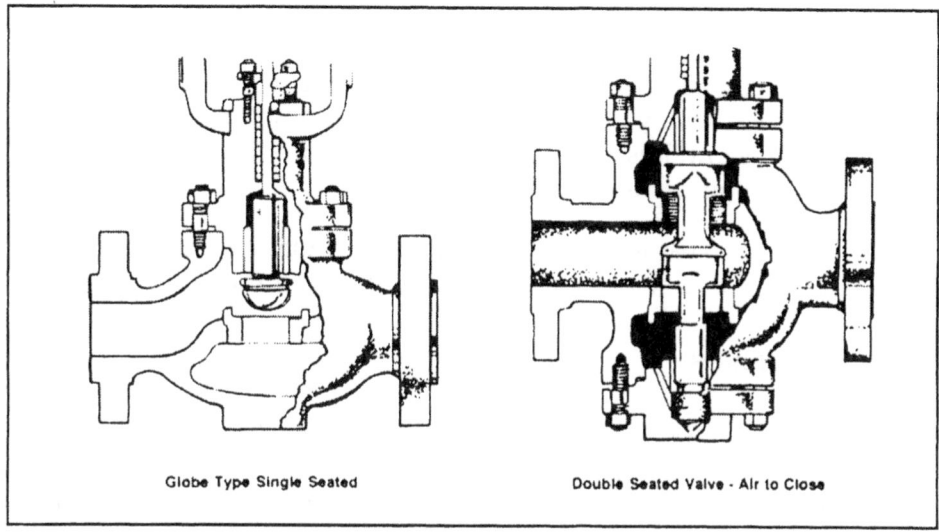

Globe Type Single Seated   Double Seated Valve - Air to Close

**Figure 7-3. Linear Stem Motion Control Valves
(courtesy of Masoneilan International Inc.)**

7-3. Rotary Control Valves

Rotary-shaft control valves have enjoyed a substantial increase in usage in recent years. Their advantages are low weight, simplicity of design, relatively high flow rates, more reliable and friction-free packing, and relatively low initial cost. They cannot usually be used in sizes below 1 inch.

The most common rotary-shaft control valve is the butterfly valve, which is illustrated in Fig. 7-4. Butterfly valves are used in sizes from 2 inches to 36 inches, or larger. They are often used in applications that involve large flows at high static pressures but have limited pressure drop availability. Properly selected, the butterfly valve offers the advantages of low cost, lightness of weight, simplicity, and space-saving size. It also exhibits good flow control characteristics.

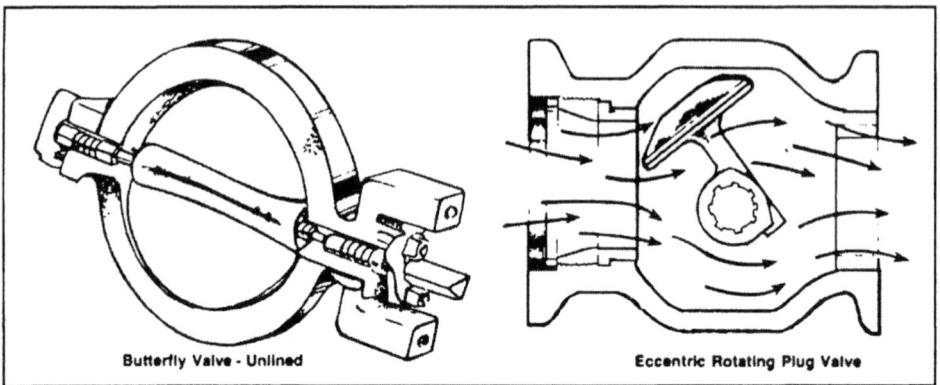

**Figure 7-4. Rotary Shaft control Valves
(courtesy of Masoneilan International Inc.)**

One type of rotary or rotating shaft control valve that is particularly useful is the eccentric cylindrical plug valve. It is actually a modification of the plug cock widely used for shutoff service. Its relative capacity is high and its cost is low. It is especially useful on services involving corrosive fluids, viscous liquids, or suspended solids.

Also illustrated in Fig. 7-4 is the eccentric rotating plug valve. This type can serve a large percentage of all process control requirements.

7-4.   Ball Valves

Ball valves are a special class of rotary valves. The use of ball valves for control purposes has grown rapidly since the early 1960s, although ball valves as such are much older. The ball valve has historically been used principally as a tight shutoff hand valve. In recent years, however, ball valves have been automated for control purposes and have shown excellent "rangeability" and excellent suitability for handling slurries.

There are two distinctly different types of ball control valves. One involves a ball (a complete sphere) that has a waterway through it. This is generally referred to as the full ball type. The second type is developed along the lines of the concentric plug valve and utilizes a hallowed out spherical segment, or partial ball, that is supported by shafts. These are referred to as "characterized" ball valves and are illustrated in Fig. 7-5.

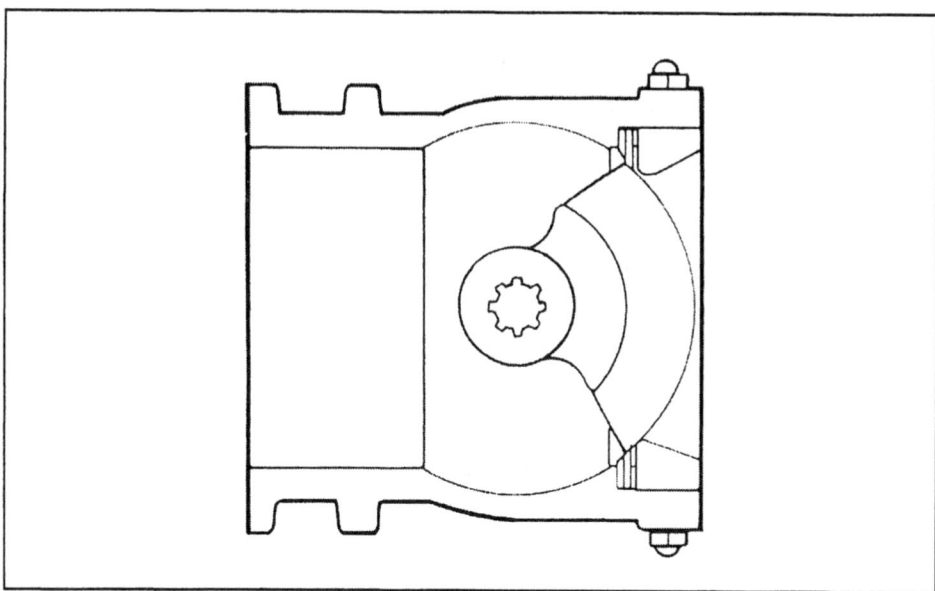

Figure 7-5. Cross Section of a Characterized Ball Control Valve (courtesy of Masoneilan International Inc.)

Both types of ball valves are quarter-turn rotary valves. The characterized ball valve is basically a simple segment of a sphere, which forms a crescent-shaped flow path that produces flows of equal-percentage characteristic for high-capacity designs and a linear characteristic for low-capacity designs. To obtain other characteristics, the leading edge of the segment can be contoured, as in the V-notch design and the parabolic port designs.

In the full ball valve design, the shape of the control orifice goes from circular to elliptical as the ball is moved from its open to its closed position. This gives an essentially equal-percentage behavior. Changing the shape of the waterway in a full ball valve is not a practical approach for changing the valve's characteristic, and when such a goal is desired it is often accomplished by using interchangeable cams in the valve positioner.

Ball valves have the highest flow capacity of any commonly used control valves. They are useful whenever slurries are involved; they also provide tight shutoff.

7-5.   Control Valve Characteristics

Control valves are characterized by their *rangeability,* their *turndown,* and their flow behavior.

Rangeability is the ratio of the maximum controllable flow through the valve to the minimum controllable flow through the valve. In typical operations, control valves do not close off entirely because to do so might damage the valve seat or cause the valve to stick. Flow in the closed position, therefore, is typically 2 percent to 4 percent of the maximum flow. This corresponds to a rangeability of fifty to twenty-five. Linear-stem motion control valves show a rangeability of twenty to seventy.

Turndown is defined as the ratio of the normal maximum flow through the valve to the minimum controllable flow. A practical rule of thumb is that a control valve should be sized so that the maximum flow under operating conditions is approximately 70 percent of the maximum possible flow. Thus, the turndown ordinarily is about 70 percent of the rangeability.

Flow through a control valve depends not only on the extent to which the valve is open, that is, on the travel, but also on the pressure drop across the valve. The next section of this unit provides specific insights into the broad question of the relationship between flow and pressure drop for a given valve opening.

7-6.   Control Valve Selection and Sizing

There are many important factors to be considered in control valve selection. These include the following:

a.   the rangeability of the process and the maximum specific range of flows required by the process

b.   the range of operating loads to be controlled

c.   the available pressure drops at the valve location at maximum flow and at minimum flow

d.   the nature and condition of the fluid flowing through the valve

Control valve rangeability must exceed the rangeability of the process by some reasonable safety margin.

Not only must control valves have sufficient rangeability, but they also must be able to moderate flows that are well below the minimum requirement of the process. In addition, the maximum flow rate through the valve must exceed the peak requirement of the process. For many process plants, the normal range of operating loads is narrow, and, in such cases, the control valves maintain very nearly constant flow rates. Valves selected to operate at 60 percent to 70 percent of full capacity are normally used in such applications. There are many process units, however, in which the operating conditions vary considerably, that is, the load varies significantly. In such conditions, equal-percentage flow characteristics are often used since they provide a more constant fractional sensitivity at all operating levels.

The nature and condition of the fluid have obvious effects on valve selection. The entire question of valve pressure drop enters not only into the sizing of the valve but also into the nature of the dynamic performance of the valve. The proper size of a control valve is most important to the overall operation of the feedback control system. If the valve is oversized, it tends to operate only slightly open, and the minimum controllable flow is too large. In addition, the lower part of the flow characteristic curve is quite often nonuniform in shape. On the other hand, if the valve is undersized, the maximum flow needed may not be obtainable.

In sizing control valves, it is standard practice to combine many terms from the basic orifice equation into the following general relationship for incompressible liquids:

$$m_L = C_v \sqrt{\frac{\Delta P}{G}}$$

where:

m_L = liquid-flow rate at the conditions where specific gravity G is taken, gallons per min (gpm)

G = specific gravity of liquid (referred to water) at either flowing or standard conditions

C_v = valve coefficient

ΔP = pressure differential, psi

The *valve coefficient* C_v is defined as the flow rate of water in gallons per minute provided by a pressure differential of 1.0 psi through a fully opened control valve. The size coefficient for any control valve must be determined by actual test. The valve coefficient for a control valve of the linear-stem travel type is very approximately equal to the square of the nominal valve size multiplied by ten.

For the flow of compressible fluids, the valve coefficient is also employed with suitable conversion factors. The form of the equation is an approximation to the complete isentropic flow equation:

$$m_g = 63.3 C_v Y \sqrt{\Delta P Y_1}$$

where:

m_g = flow rate of gas, pounds per hour
C_v = valve coefficient (obtained for liquids)
ΔP = pressure differential across valve, psi
Y_1 = specific weight, lb/ft^3, at upstream conditions
Y = expansion factor, ratio of flow coefficient for a gas to that for a liquid at the same Reynolds number, varies from 0.667 to 1.0

The entire question of determining control valve size, flow rate, or pressure differential is made in most industrial applications by using special-purpose calculators or special-purpose computer programs. In actual practice, the sizing of control valves is more complex than is illustrated here. The equations we have shown here do illustrate the basic principles involved, however. For specific details on the sizing of valves, the student should consult the books written on the subject or control valve vendors.

7-7. Control Valve Dynamic Performance

It is most important that you gain a specific understanding of the dynamic characteristics of a control valve installed as a part of a total piping network. The concepts involved significantly affect the valves' operating characteristics and the total performance of the feedback control system.

A piping network is shown in Fig. 7-6. In this case, a centrifugal pump takes a suction on a feed tank and pumps a liquid stream through a preheater and into a distillation column. There is a liquid-flow control loop that serves to control the actual liquid-flow rate into the distillation column.

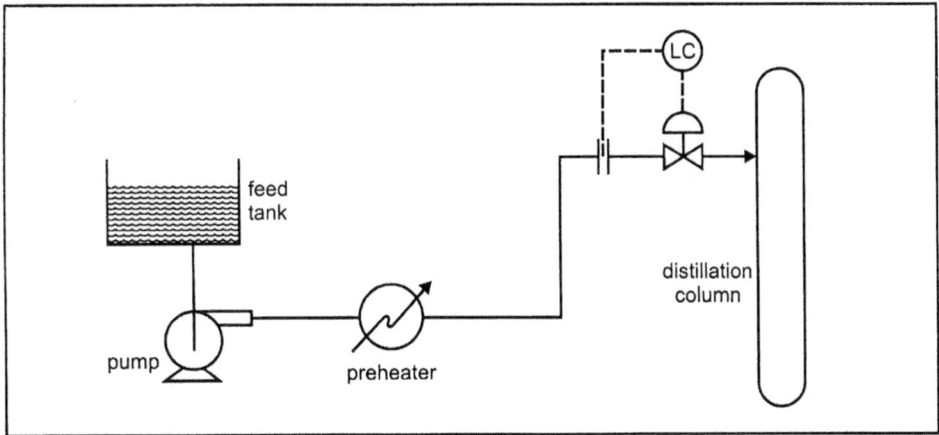

Figure 7-6. Example Piping Network

If you closely inspect the total distribution of head loss throughout this piping network, you will get some insight into the dynamic characteristics of the valve. The centrifugal pump might have a total characteristic head curve like that shown in Fig. 7-7.

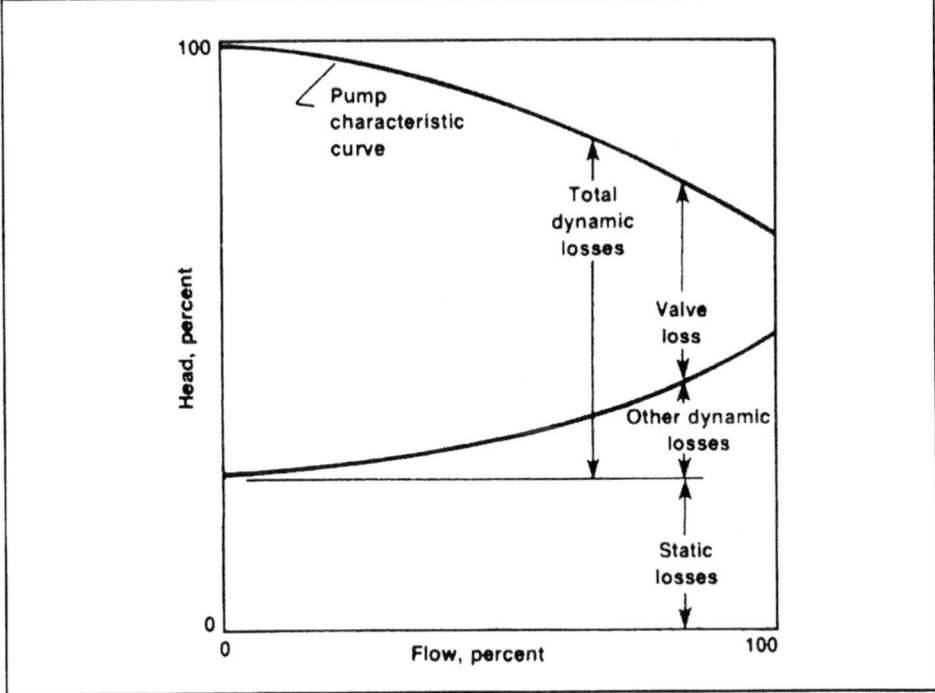

Figure 7-7. Head Distribution in the Network

This pump output head is "consumed" by a number of different types of pressure drops. Many of these are static in nature and independent of the amount of fluid flowing through the piping network. One example of such static loss is the potential energy that is required to raise the liquid to the feedpoint on the column. A second static loss is illustrated by the operating pressure of the column itself, which must be overcome by the inlet liquid feed.

In addition to these static losses, there are many *dynamic losses* involved in the piping network besides the losses associated with the control valve. These additional dynamic losses tend to increase as the square of the flow rate through the piping network. These losses are the friction losses in the piping, flow through the preheater, and flow through the orifice plate. After the static and dynamic losses, all of the other head losses of the flowing fluid must be consumed across the control valve.

In a very real sense, the control valve is a bottleneck but one that is designed and installed on purpose. If it is not the bottleneck, it is not "in control." As flow increases and a smaller percentage of the total dynamic drop occurs across the valve, then the piping network and not the valve tends to become the bottleneck.

When most valves are sold, the customer obtains a single characteristic curve for the valve. This is typically a curve like that in Fig. 7-2, which illustrated the valve flow characteristics of the three basic types of sliding-stem control valves. These basic curves are representative of the situation in which the total dynamic drop occurs across the control valve.

In actuality, for every single control valve there is a whole family of control valve sensitivity curves. The plotting parameter in the family of such curves is the percentage of the total dynamic drop that occurs across the control valve. This is illustrated in Fig. 7-8, which shows a family of curves for a typical equal-percentage valve and for a typical linear valve. In this illustration, $\%\Delta P_v$ is the "Valve loss," which was defined in Fig. 7-7 as a percentage of the "Total dynamic losses."

Note that the characteristic sensitivity of the valve may be significantly altered if you have less than the total dynamic drop occurring across the control valve. These questions have a remarkable impact on the tuning of process control systems, a topic we will discuss in Unit 9.

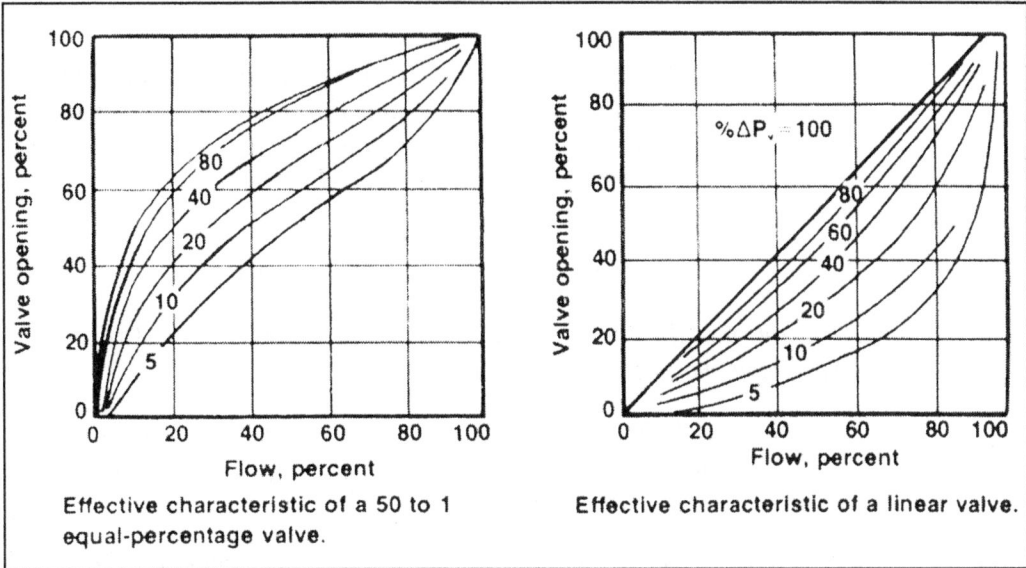

Figure 7-8. The Effect of Percentage of Dynamic Drop across the Control Valve

7-8.   Power Failure

It is inevitable that power will fail occasionally in a process operating unit, and this produces a power failure for the final control element. Whether the final control element is pneumatic, hydraulic, or electric will affect some of the individual aspects of the emergency. In general, there are three possible patterns that might be encountered in power failure:

    a.   The final control element may fail open.

    b.   The element may fail "last position."

    c.   The element may fail closed.

In any given instance, it is possible that any one of these three situations may be desirable. The practitioner must perform significant process analysis to make specific decisions about the way the failure mode should be designed; it is an important part of the total process control strategy.

Further Study

    1.   H. D. Bauman, *Control Valve Primer: A User's Guide*, 3d ed. (ISA, 1998).

    2.   Guy Borden Jr., *Control Valves*, Practical Guide Series (ISA, 1998).

EXERCISES

7-1.   *An equal-percentage control valve has 10 gpm flowing through it. When the air-to-valve signal increases 0.5 psi, flow through the valve increases 1.5 gpm. Much later, the base flow through the valve is operating at 20 gpm when the air-to-valve signal increases 0.5 psi. How much will flow increase through the valve?*

7-2.   *Repeat Exercise 7-1 but assume the valve has linear trim.*

7-3.   *A control valve flowing water has C_v of 20 and an assigned pressure drop of 10 psi. How many gpm will the valve flow?*

7-4.   *A control valve is flowing air. It has a C_v of 32, an assigned pressure drop of 10 psi, an expansion factor of 0.8, and a specific weight of 3.103 lb/ft^3. Calculate flow through the valve.*

7-5.   *Given an equal-percentage valve installed with 10 percent of the dynamic pressure drop across the valve, assume the valve operates about 50 percent to 70 percent open under normal conditions. Does the valve act more like an equal-percentage valve or more like a linear valve?*

7-6.   *Consider several actual feedback loops either at home or at work, such as, the heating system, the oven, and the air compressor controller at your service station. Decide for each loop how you would like the final element to "fail" in the event of power loss.*

Unit 8:
Process Dynamics

UNIT 8

Process Dynamics

To get a good grasp of the fundamentals of process control theory, it is important that you develop an general appreciation of process dynamics. This unit introduces many of the general concepts of process dynamics and much of the associated terminology.

Learning Objectives — When you have completed this unit, you should:

A.   know the general response characteristics of a first-order lag component that has been subjected to a step input

B.   be able to determine graphically a time constant for a first-order lag system that has been driven by a step input

C.   be able to identify a dead time on a process response curve

D.   understand the effects of process lags and dead times on loop process dynamics

8-1.   First-Order Lags

The first-order lag is the most common type of dynamic component encountered in process control. To study it, it is helpful to look at response curves when the component under study is subjected to a step input such as this:

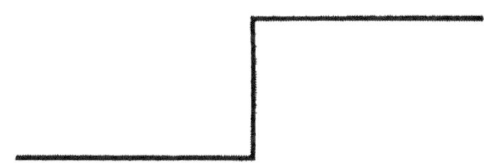

The advantage of using such a step input as a forcing function or driving force is that the input is at steady state for the time previous to the change, and then the input is instantaneously switched to a new steady-state value. When the resulting output or response curve is inspected, you then observe the transition of the component as it passes from one steady state to a new steady state.

For a first-order lag component, the response to a step input is as illustrated in Fig. 8-1.

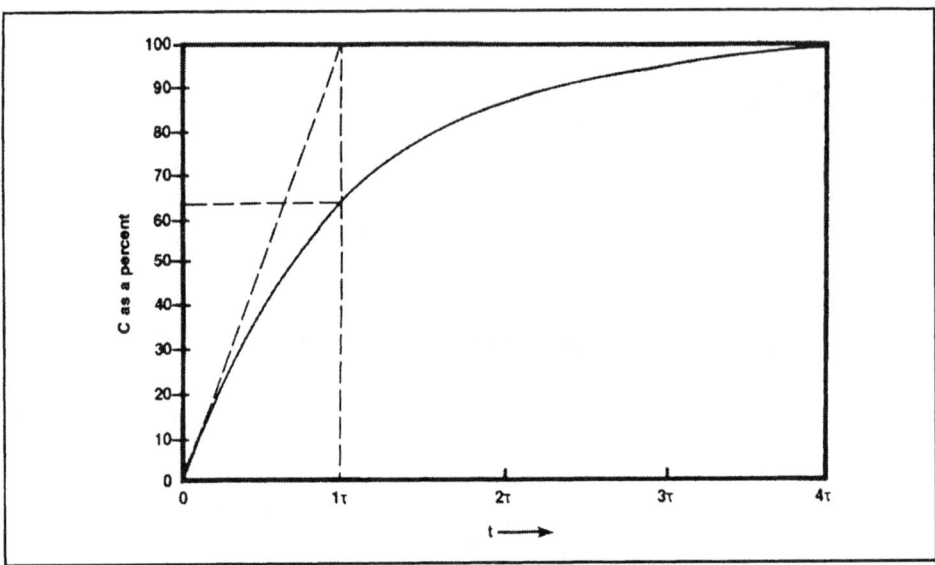

Figure 8-1. The Response of a First-Order Lab Component to a Step Input

The component whose response is shown in Fig. 8-1 is called a first-order lag because the output lags behind the input, and the differential equation that describes the underlying mathematics is a linear first-order differential equation.[1] This is the most common type of dynamic component encountered in practice. In addition to being referred to as first-order lags, these components are often referred to as *linear lags* or *exponential transfer lags*. These components are characterized by their capacity to store material or energy, and the dynamic shape of these response curves is described by a *time constant*. The time constant τ is meaningful both in a physical sense and in a mathematical sense. In a mathematical sense, it represents, at any moment, the future time necessary to experience 63.2 percent of the change remaining to occur. This assumes that a step input has been introduced to the component under study. This is shown in Fig. 8-1.

If you look closely at the first-order response curve in Fig. 8-1, you also see that the response is always falling off. The rate of response is a maximum in the beginning and is continuously decreasing from that time onward. If

---

1. The actual equation is as follows:

$$\tau\frac{dc}{dt} + c = Kr$$

where c = output, r = input, K = gain, τ = time constant. In block diagram form, this is

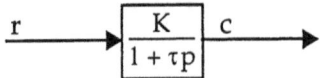

where p is the Heaviside operator d/dt.

the system were to continue to change at its maximum response rate (that, of course, occurs at the origin), it would reach its final value (or 100 percent) in one time constant.

Table 8-1 gives the numerical value of the changes that take place in a first-order lag response to a step input. In the first time constant of elapsed time, 63.2 percent of the total response will occur. Within the next time constant, 63.2 percent of the remaining 36.8 percent of the change will take place. This repeats itself without mathematical limit since, theoretically, the response never reaches 100 percent, but it does approach it asymptotically.

Elapsed Time	Percent of Total Response	Response Remaining	63.2 Percent of Response Remaining
1τ	63.2	36.8	23.2
2τ	86.4	13.6	8.6
3τ	95.0	5.0	3.16
4τ	98.16	1.84	1.16
5τ	99.32	0.68	0.429

Table 8-1. Response of a First-Order Lag to a Step Change

The time constant gives dynamic insight into the rapidity of the system or component response. It is a specific measure of the rapidity of response. Fig. 8-2 shows six different response curves, all for first-order lags, but each of the six has a different time constant. As you can see, as the time constant gets larger, the response gets slower.

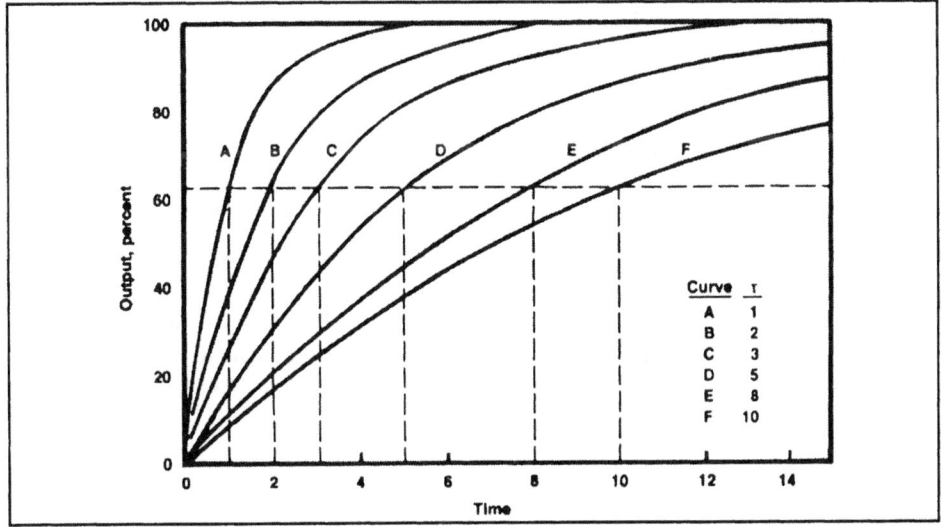

Figure 8-2. Some First-Order Lag Responses to a Step Input

8-2.   Time Constants

Frequent use is made of the time-constant concept in process control theory. It is an idea that is not entirely new to the technical person since a good basis for understanding it can be found in simple electrical circuits. The following is a simple electrical RC network:

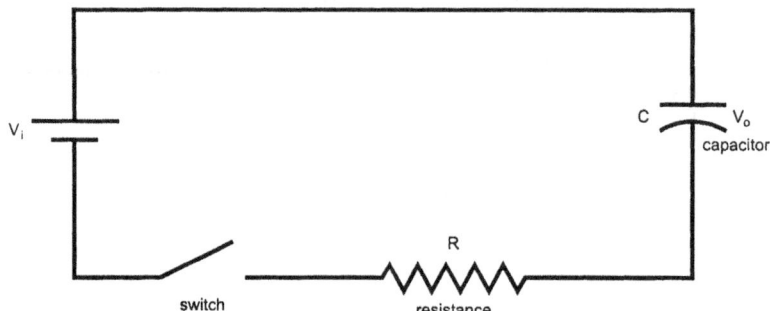

Assume that we wish to derive the equation to describe the charging of this capacitor when a voltage V_i is suddenly applied to the circuit (by closing the switch). V_o is the voltage across the capacitor, and it turns out that the buildup of V_o is given by a linear first-order lag.[2] The product RC, or resistance times capacitance, is the natural time constant for this simple network. If one checks the units involved, the product of RC has the units of time.

The product of resistance times capacitance is the system time constant for the simple electrical network just shown, but this physical analogy can also be used to gain insight into time constants of systems other than electrical networks. To do so, it is advantageous to think of the reciprocal of resistance, which is termed "conductance." For many physical systems, time constants may be viewed more easily as capacitance divided by conductance:

time constant = (resistance)(capacitance) = capacitance/conductance

Table 8-2 gives a broader view of the time constant analogy between various physical systems. In an earlier example (in Unit 4), you observed the first-order lag response of a bare bulb thermometer immersed in an agitated bath (a step input). In that case, the capacitance for the bulb is the mass of bulb material (that must change in temperature) times its weighted average heat capacity. The conductance is the heat transfer

---

2.   The actual equation is as follows:

$$RC\frac{dV_o}{dt} + V_o = V_i$$

Variable	Electrical	Liquid Level	Thermal	Pressure
Quantity	Coulomb	Cubic foot	Btu	Cubic foot
Potential or force	Volt	Foot	Degree	$\dfrac{\text{Pounds}}{\text{Square inch}}$
Flow rate	$\dfrac{\text{Coulombs}}{\text{Second}}$ = Amperes	$\dfrac{\text{Cubic feet}}{\text{Minute}}$	$\dfrac{\text{Btu}}{\text{Minute}}$	$\dfrac{\text{Cubic feet}}{\text{Minute}}$
Resistance	$\dfrac{\text{Volts}}{\text{Coulombs}}$ per second = ohms	$\dfrac{\text{Feet}}{\text{Cubic feet}}$ per minute	$\dfrac{\text{Degrees}}{\text{Btu per}}$ minute	Pounds per square inch $\dfrac{}{\text{Cubic feet}}$ per minute
Capacitance	$\dfrac{\text{Coulombs}}{\text{Volts}}$ = farads	$\dfrac{\text{Cubic feet}}{\text{Foot}}$	$\dfrac{\text{Btu}}{\text{Degree}}$	$\dfrac{\text{Cubic feet}}{\text{Pounds per}}$ square inch
Time	Seconds	Minutes	Minutes	Minutes

Table 8-2. Time Constant Analogy between Basic Physical Systems

coefficient between the bath liquid and the bulb times the heat transfer area. If this capacitance is divided by this conductance, you find the natural time constant for the bulb expressed in units of time.

It is important that you begin to get a general feel for the physical meaning of time constants. With practice, you can inspect an individual component or process and achieve some appreciation of its capacity to store material or energy as well as a sense for the conductance of material or energy into or through the component or process. In this way, you will achieve insight into the time constants involved, and, therefore, you will gain a better understanding of the process dynamics of an individual component or process. Such understanding will materially assist you in process control applications.

8-3.   Higher-Order Lags

Of course, not all process dynamics can be neatly characterized by first-order lags. Often a situation is encountered where a step input will produce a response curve such as the following:

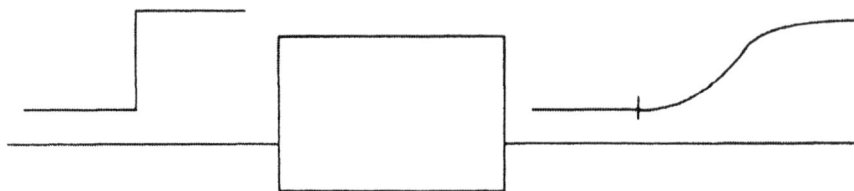

Note that the maximum rate of change for the output response curve does not occur at the origin but at some later time. This implies that a higher-order system is involved and that a first-order lag is inadequate to characterize the system. Higher-order systems can be the result of several different situations:

a. several first-order lag processes may be encountered in series.

b. the installed feedback controller may introduce a characteristic differential equation that, when considered in series with the other system components, makes the overall description of the system higher ordered.

c. mechanical or fluid components of the system may be subject to accelerations, that is, to inertial effects (usually this is a minor possibility).

d. the process may be a distributed process that gives a response curve that can be described only by higher-order differential equations or by partial differential equations.

When first-order lags are encountered in series, the way they are connected in series will have a lot to do with the resulting process dynamics. They may be connected so as to produce *independent stages* or *interacting stages*; these are illustrated in Fig. 8-3. For example, both of the tanks in Fig. 8-3 taken separately would give first-order lag responses to a change in inlet flow rate. But the manner in which they are connected causes their combined response to differ between case (a) (independent stages) and case (b) (interacting stages).

In the (a) case, the height of the liquid in the second tank does not influence the flow of liquid out of the first tank, but in the (b) case, it is clear that the height of liquid in the second tank will influence the rate at which liquid flows out of the first tank. When stages are interacting with one another, the resulting process response will always be more sluggish than would be expected from simple first-order lags connected as independent stages. This case is of special importance in thermal systems because they are usually sluggish.

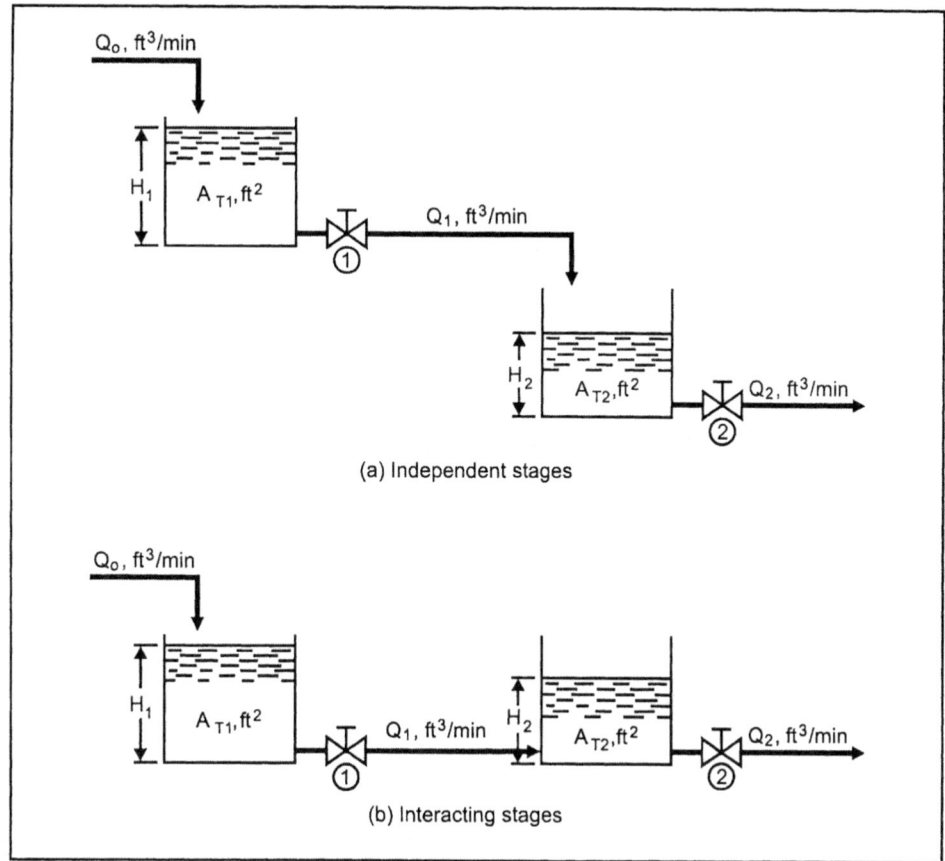

(a) Independent stages

(b) Interacting stages

Figure 8-3. Independent and Interacting Stages

Higher-order response, in which the maximum change in process response rate does not occur at the origin, is quite common. Fig. 8-4 shows some typical effects of thermowells on thermometer bulb response. Curve A is a plain bare bulb with first-order lag dynamics. As the bulb is provided with more protection, the response gets more and more sluggish, is higher ordered, and exhibits the S-shaped response curve of higher-order systems. In these cases, there is, of course, much more lag involved, and the process becomes more difficult to control.

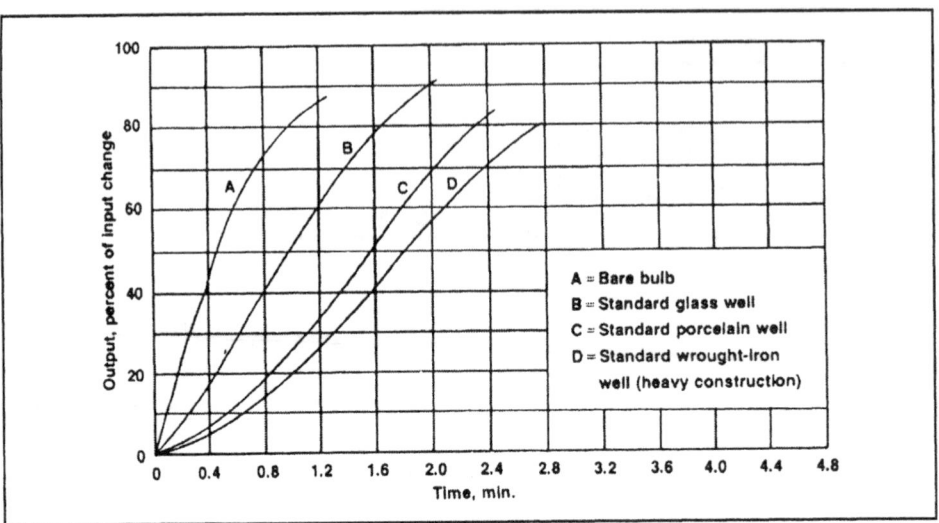

Figure 8-4. Some Effects of Thermowells

8-4.   Dead Time

In process dynamics you often encounter a process response curve in which there is a "dead time" before there is any dynamic response whatsoever. This might appear as follows:

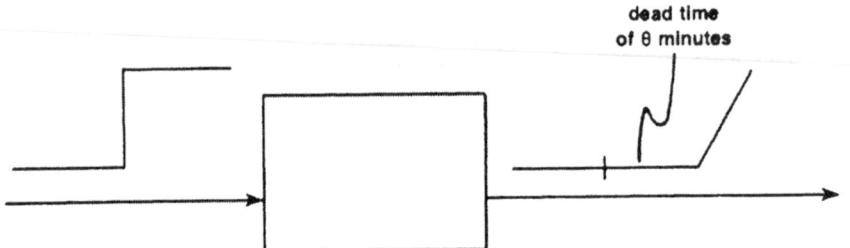

Such dead times, shown here as θ, are some of the most difficult situations to control. The difficulty stems from the fact that during this time period (the dead time) there is absolutely no response whatsoever, and, therefore, there is no information available to initiate corrective action.

For some specific insights into how a dead time might occur in practice, refer to Fig. 8-5, which illustrates simple feedback control of a system where steam is injected into a water tank to produce hot water. There is a thermobulb in the outlet stream to measure the temperature of the exiting hot water. The control system increases or decreases the manipulated variable, that is, the steam flow, to produce the outlet water at the desired temperature.

A difficult question arises when one begins to debate where this temperature bulb should be located: Where should the feedback sensor be installed? It is tempting to say that it should be installed further down the outer pipe so it is closer to the location where the water will actually be used. This rationale sounds appealing because it is at the point of usage that the temperature of the water is significant; that is, it is of little importance what the water temperature might be in the tank itself or at the tank exit. This type of reasoning, while appealing, is disastrous.

As the temperature bulb is moved farther and farther down the outlet pipe, a dead time is actually installed in the feedback loop before the sensor. In this case, the dead time is equal to the distance down the pipe to the point where the bulb is installed, divided by the velocity of the water inside the pipe. As a result of this encounter with such a dead time in the feedback loop, overall loop control deteriorates, and it may reach a point where it is impossible to achieve any feedback control at all.

Often, dead times are referred to by other names such as *distance-velocity lags* or *transportation lags* or *time delays*. These other descriptive names can be appreciated very specifically in terms of the example shown in Fig. 8-5.

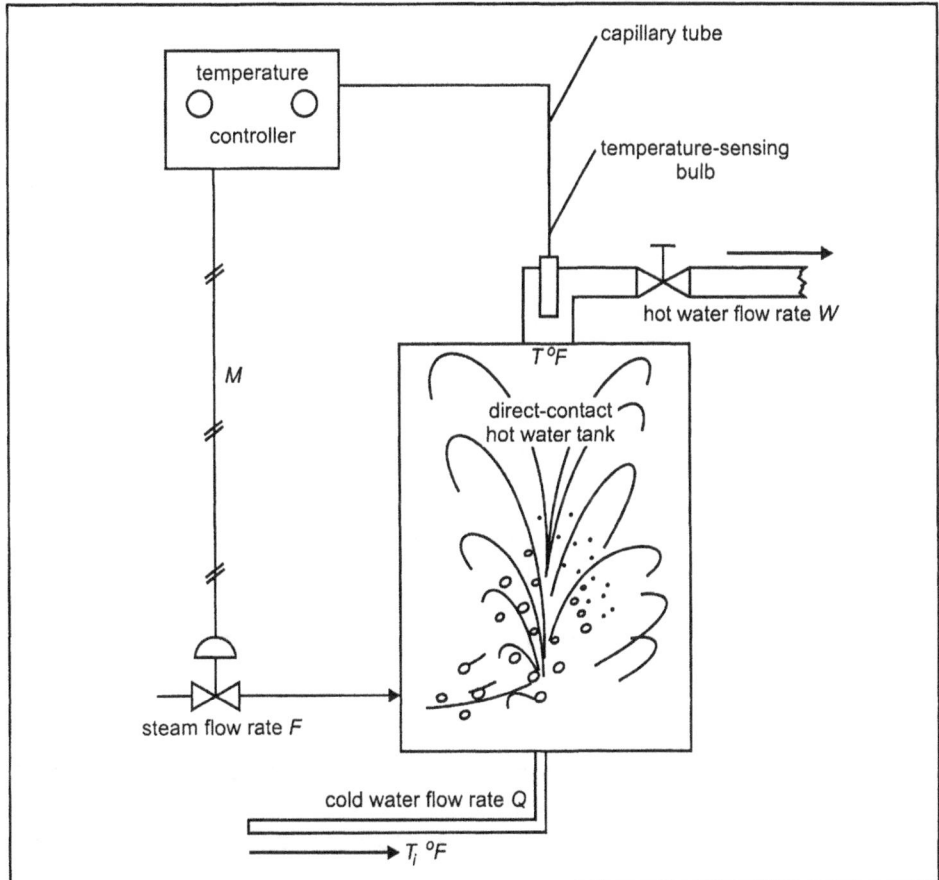

Figure 8-5. Sample System

Quite often, dead times are unavoidable in actual practice, for example, in a rotating kiln, in a plug flow reactor, or in similar situations. However, there are other cases where the designer's lack of insight and forethought cause the sensor to be installed in an inappropriate location, and for no good reason whatsoever a dead time is inadvertently placed into the feedback control loop. Such mistakes cause serious operating problems and are a severe indictment of the designer. Sometimes such problems occur as a result of the type of reasoning we discussed for the hot-water tank example in Fig. 8-5. Sometimes such problems are a result of a desire to place a sensor in a spot where maintenance might be achieved more easily. Whatever the reason, dead times should be avoided wherever possible. They are easily the most difficult dynamic elements to control.

Dead times can occur in combination with nondynamic components:

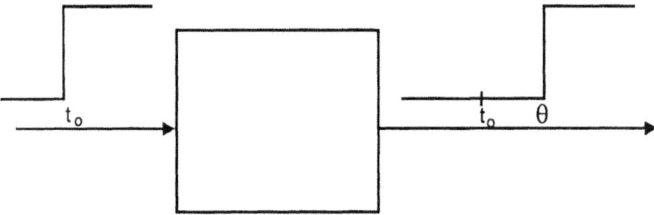

Or they may occur with first-order lags:

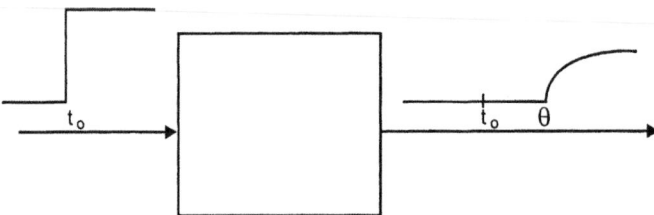

Or they may be found in combination with higher-order lags:

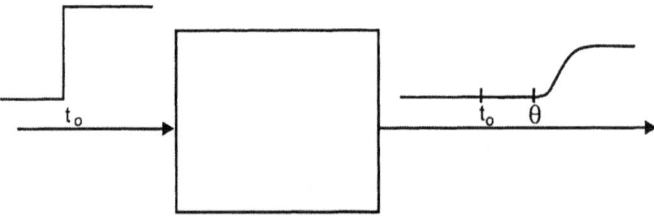

Whenever they are encountered, you should be able to simply pick them off the process response curve by simply measuring the amount of time that transpires before any output response occurs after an input to the system.

8-5.   Closed-Loop Response versus Open-Loop Response

In the examples we have previously presented, we devoted attention to the "open-loop" response curve. The implication was that the feedback loop was not closed and that you were looking at the open-loop behavior of individual components or systems.   However, the effective dynamic behavior of a component in a feedback loop will be much different when that loop is closed. This fact cannot be overemphasized.

One generalization needs to be made explicitly clear. When a loop is closed around a dynamic component, generally it will become less stable, that is, it will respond faster and more dynamically. This is a fact that all process operators realize very quickly. When they encounter dynamic control problems with a feedback loop, they immediately place the controller on "manual." This means that they break the feedback loop and operate the control valve directly.

In process control applications it is rare to encounter any open-loop oscillations. But if you take a system with several dynamic components, place these components together in a loop, and then close the loop, it is quite possible for the system to have an output response curve that exhibits oscillations. In order for us to discuss this more fully, you must become familiar with some of the descriptive terms used in characterizing systems that exhibit oscillation. These descriptive terms are illustrated in Fig. 8-6.

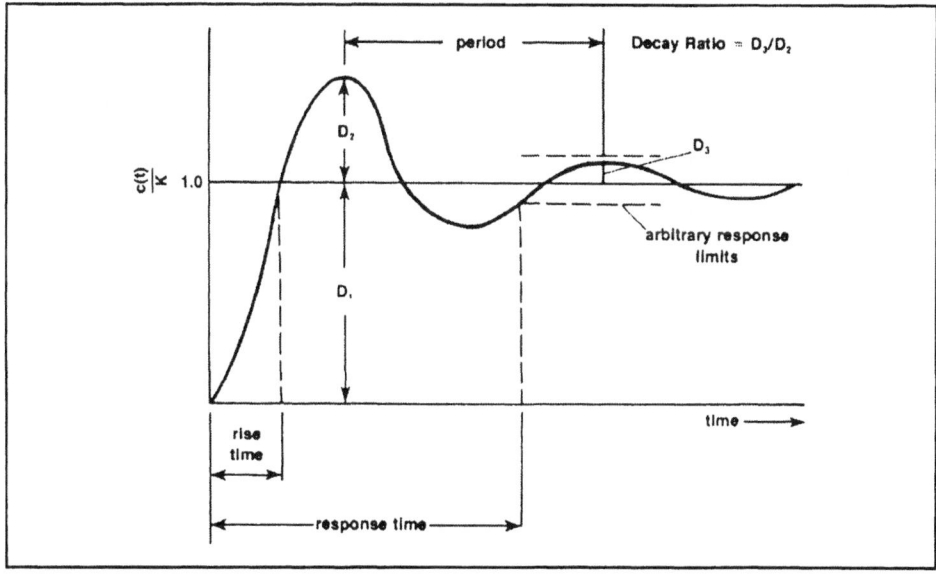

Figure 8-6. Descriptive Terms for Underdamped Response Curves

To illustrate the effects of closing a process control loop, inspect the two response curves shown in Fig. 8-7. Assume there are two first-order lags with an overall gain of five and that these first-order lags have time constants equal to one another and, for illustrative purposes, equal to one. The open-loop response is shown in Fig. 8-7. When the loop is closed and subjected to the same step input, the response curve is as shown in the figure. Notice the dramatic change. This illustrates the need to make significant differentiations between open-loop and closed-loop response considerations.

The point has been made *that a loop is less stable when it is closed*. This generalization can be carried a step further. If you increase the gain on a loop that has been closed, the loop progressively becomes less and less stable. It exhibits more and more oscillatory behavior. In Fig. 8-7, if the gain on the loop had been higher, when the loop was closed the oscillation shown in the closed-loop response curve would have been larger. As you increase the gain higher and higher, you speed up the response of the loop. This is desirable because it speeds the controlled variable's transition as a

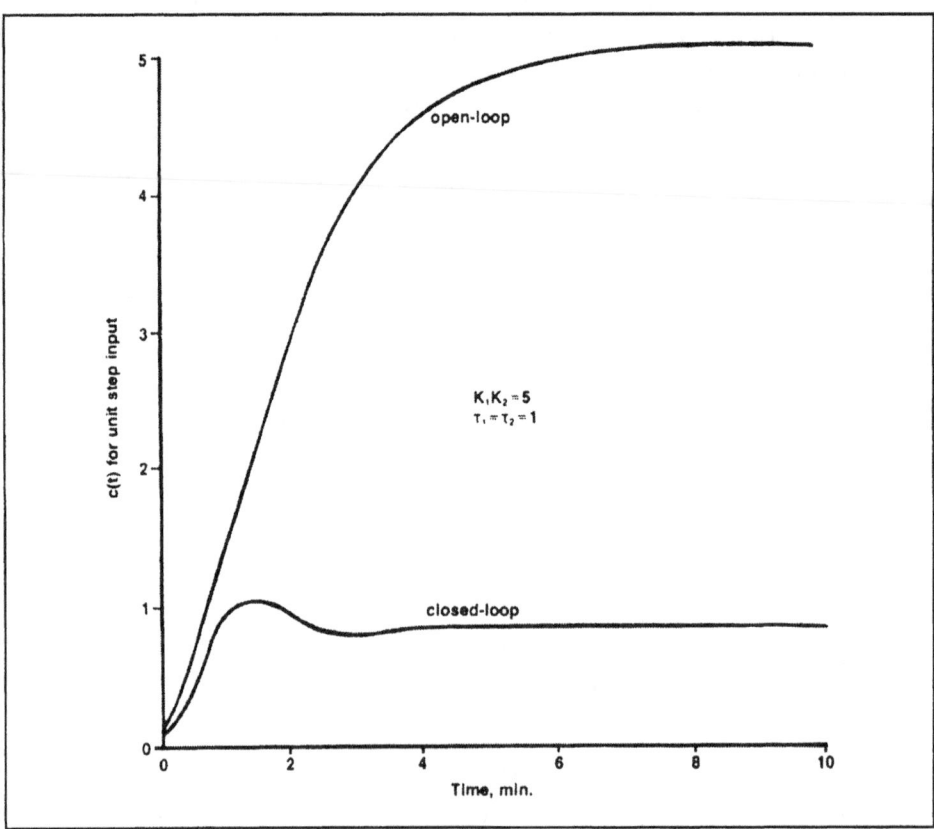

Figure 8-7. Open-Loop versus Closed-Loop Response

result of set point changes and gives quicker response to disturbances. But as the gain is increased higher and higher, the loop becomes less and less stable. Therefore, there must be a trade-off between the two phenomena. In effect, the *optimum gain* is the one that gives the proper trade-off between speed of response and stability for the loop.

8-6.   Some Generalizations

It is especially important to understand some significant generalizations that can be made about process dynamics. In the last section, we demonstrated that when loops are closed they become less stable, that is, they show more significant dynamic changes. Also, as the gain of a loop increases, it becomes less stable—but at the same time it responds faster.

In general, the more lag—that is, the more time constants and the more dead times—that are encountered around a process loop, the worse the control problem. The particular order or sequence of the time constants or dead times is not especially significant. It does not make any difference what particular piece of hardware is involved. That is, it does not make any difference whether a particular time constant is associated with the valve, with the sensor, or with the process itself. No matter where it occurs, a time constant or dead time has the same deleterious effect on the overall control of the system.

These generalizations concerning process time constants include the controller. In Unit 6 we showed that the addition of the reset mode to the controller also introduces some lag in the controller, and, as a result, the loop is less stable when reset is added. The desirability of eliminating offset is often a justification for reset, but you must appreciate the penalties that are involved in using it.

There are cases where the process may be so rapid that it, in effect, needs to be filtered or slowed down in order to increase controllability. There is one particularly important case of this: liquid-flow control loops. In these, the process itself has almost instantaneous response, and, as a result, there will be a lot of high-frequency noise transmitted around the loop. This is undesirable, and to counteract this the reset mode can be added to the flow controller to provide some lag or some filtering to the loop dynamics. In effect, lag is introduced via the reset mode in the controller. Of course, such an addition inadvertently eliminates offset. An alternative way to reduce this noise problem is to operate the loop at a low gain (high-proportional band.)

The point is that the more process lag--that is, the more time constants and dead times around a loop--the worse the control situation (except as we noted in the previous paragraph). In addition, the more this process lag is distributed around the loop, the worse the control case becomes. To say this differently, the more the process lag is concentrated in a specific component, the better the control situation. For example, a first-order lag with a time constant of ten minutes would not be too difficult to control. If, however, this same process lag were broken into two time constants of five minutes each, the control would be more difficult. If it were broken into ten time constants of one minute each, it would be even worse. As a matter of fact, if it were broken into a large number, for example, two hundred time constants of a few seconds each—but whose total was ten minutes— then the behavior would approach that of a ten-minute dead time (the most difficult case of all).

Little can be done about process dynamics in many situations. The dynamic behavior of the process usually is established by the function it serves. It is expected that the practitioner will design and install a transmission system that operates sufficiently fast so that it does not contribute significant lag to the overall process control loop. This leaves the control valve and sensor where the practitioner can make choices that have dynamic implications. Sometimes these choices are made without explicit consideration of the inherent dynamics of the sensor and/or control valve. The dynamic behavior of these two components is especially important, and many control loops are made unnecessarily difficult to operate simply because of design problems. The practitioner installs the sensor in such a way that it exhibits very slow and sluggish behavior, or he or she installs the control valve in such a way that its lag is a significant contributing factor to the poor performance of the loop. The goal of the practitioner should be to have a good appreciation of the dynamic behavior of all the hardware he or she designs and installs.

EXERCISES

8-1. *Estimate the time constants for the three components whose step-response curves are shown here:*

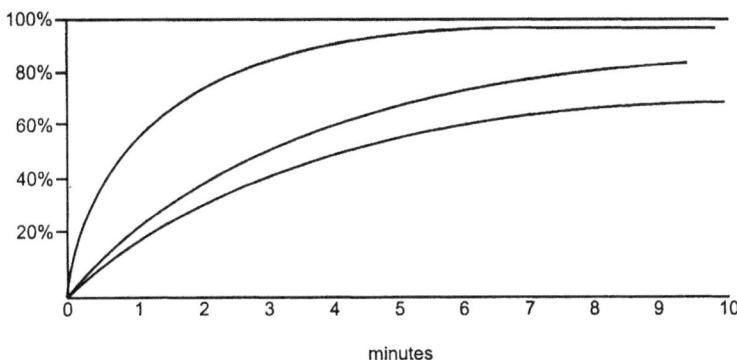

8-2. *What can you say about the dynamics of a first-order lag type component that has a time constant of zero? Of infinity?*

8-3. *A stirred tank has a volume of 28 ft³ and a flow rate through it of 6 ft³/min. What is its time constant?*

8-4. *Consider the bare thermometer bulb of Fig. 4-1. The bulb has a mass of 0.26 lb and a heat capacity of 0.6 Btu/(lb * °F). Heat transfer occurs through a bulb surface area of 0.16 ft², which has a surface film coefficient or conductance of 4.4 Btu/(min * ft² * °F). What is the bulb's time constant?*

8-5. *How much dead time is there in the component whose step response is shown here:*

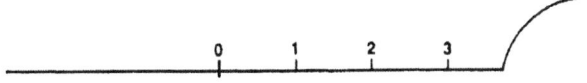

8-6. *Consider Fig. 8-5. If the bulb were installed 16 ft down the exit pipe and the flow velocity in the pipe were 21 ft/min, how much dead time would there be in the feedback loop?*

8-7. *Explain what happens in a feedback loop as you increase the controller gain higher and higher.*

8-8. *Explain what happens in a feedback loop as you introduce more and more lag—for example, dead time—into the loop.*

Unit 9:
Tuning Control Systems

UNIT 9

Tuning Control Systems

A feedback control system is of little value if it is improperly tuned. In this sense, it is very much like an improperly tuned automobile. It is important that you understand how the controller in a feedback control system should be tuned.

Learning Objectives — When you have completed this unit, you should:

A. have developed insight into the fundamental concepts of tuning feedback controllers

B. be able to calculate the tuning parameters for a feedback controller using the Ziegler-Nichols ultimate method

C. be able to calculate the tuning parameters for a feedback controller using the process reaction curve method

9-1.   What Is Good Control?

To tune a controller you need to determine the optimum values of the controller gain K_c (or proportional band PB), the reset time T_i (or the reset rate as repeats per minute), and the derivative time T_d. How to adjust these tuning parameters on feedback controllers is one of the least understood and most poorly practiced—yet extremely important—aspects of automatic control theory.

The first problem you encounter when tuning controllers is determining what is *good* control, and, as might be expected, it does differ from one process to the next. The most common criterion for determining good control is to adjust the controller so that the system's response curve has an amplitude ratio or *decay ratio* of one-quarter. A decay ratio of one-quarter means that the ratio of the overshoot of the first peak in the process response curve to the overshoot of the second peak is four to one. This is illustrated in Fig. 9-1.

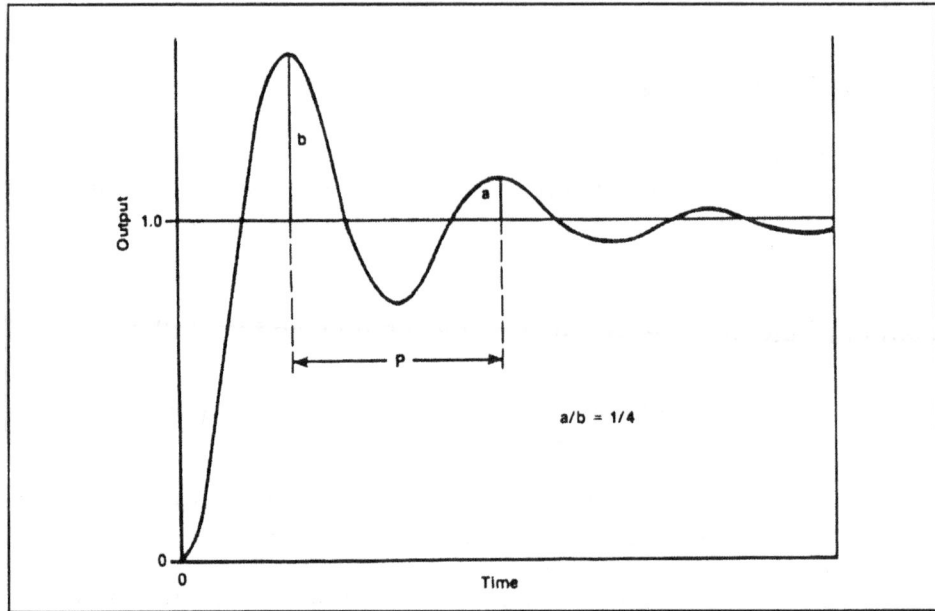

Figure 9-1. Process Response Curve for a One-quarter Decay Ratio

There is no direct mathematical justification for requiring a decay ratio of one-quarter, but it represents a compromise between a rapid initial response and a fast line-out time. In many cases, this criterion is not sufficient to specify a unique combination of controller settings. For example, in two-mode or three-mode controllers there are an infinite number of settings that will yield a decay ratio of one-quarter, each with a different period. This illustrates the difficulty of defining what constitutes good control.

In some cases, it is important to tune the loop so there is no overshoot. In other cases, you will want a slow and smooth response. Some loops need fast response, and significant oscillations are no problem. And on and on. The point is: *You* must determine what control is good for each specific loop.

9-2.   The Tuning Concept

The feedback controller is only one piece of hardware in the entire loop, and there are many other hardware items that are connected to form the balance of the loop. For the purposes of adjusting the feedback controller, it is convenient and sufficient to view everything else within the feedback loop as being "one big lump." Actually, this is the way the feedback controller sees the balance of the loop (see Fig. 9-2).

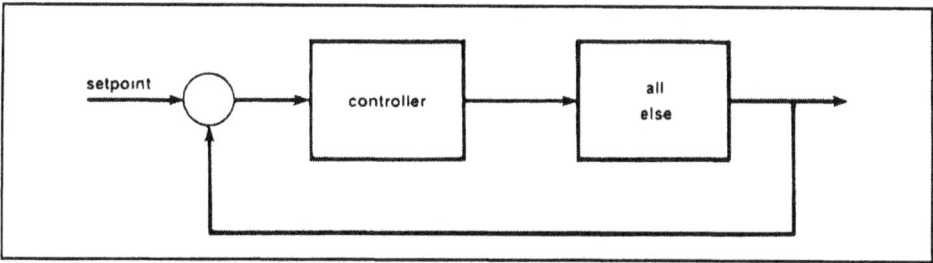

Figure 9-2. The Balance of the Loop as "Seen" by the Feedback Controller

There is one parameter to adjust in a single-mode controller--for example, the gain K_c in a proportional-only instrument. There are two parameters to adjust in a two-mode controller--for example, the proportional gain K_c and the reset time T_i in a PI controller. There are three parameters to adjust in a three-mode controller--for example, the controller gain K_c for the proportional mode, the reset time T_i for the integral mode, and the T_d for the derivative mode.[1] When you adjust the controller, the gains around the loop will tend to dictate what should be the optimum gain in the controller. Similarly, the time constants and dead times that characterize the lag dynamics for everything else around the loop will tend to dictate what will be the optimum value of the reset time and what the derivative time in the controller should be. Stating this differently, before you can calculate or select the best values for the tuning parameters in the controller, you must get some quantitative information about the overall gain and the process lags that are present in the balance of the feedback loop. This illustrates quite clearly why controllers cannot be preset at the factory but, instead, must be individually tuned for individual loops.

It is also helpful to establish a mathematical model of a process control loop and to appreciate the role of the controller as a mathematical equation within that model. As you saw in Units 3 and 8, each of the individual blocks around the feedback loop represents an algebraic or differential equation. There is a mathematical statement for each particular piece of hardware. Since all of these blocks are coupled together, all of these equations represent a simultaneous set of mathematical equations. If you are able to determine what you think good control should be, then you can, in effect, specify the overall solution to this simultaneous set of equations. Basically, the tuning of the controller represents the adjustment of the individual parameters in the equation that represents the controller. As you adjust these parameters for the controller equation, you modify the solution for the simultaneous set. You change the response of the overall system toward a point that represents good responses or good control.

---

1.   Murrill's Law states that the difficulty of tuning a controller increases with the square of the number of modes present. Humor aside, it tends to be accurate!

9-3.   Closed-Loop Tuning Methods

Techniques for adjusting controllers may be classified as either open-loop or closed-loop methods. One of the first methods proposed for tuning feedback controllers was the *ultimate method* proposed by Ziegler and Nichols in 1942. The term *ultimate* was attached to this method because to use it you had to determine the *ultimate gain* (sensitivity) and *ultimate period* for the loop. The ultimate gain is the maximum allowable value of gain (for a controller with only a proportional mode in operation) where the closed-loop system is stable.

For any feedback control system, if the loop is closed (if the controller is on automatic) you can increase the controller gain, and as you do so the loop will tend to oscillate more and more. As you continue to increase the gain further, you will observe continuous cycling or continuous oscillation in the controlled variable. This is the maximum gain at which the system may be operated before it becomes unstable; this is the ultimate gain. The period of these sustained oscillations is the ultimate period. If you increase the gain further, the system will become unstable. Ultimate gain and ultimate period are illustrated in Fig. 9-3.

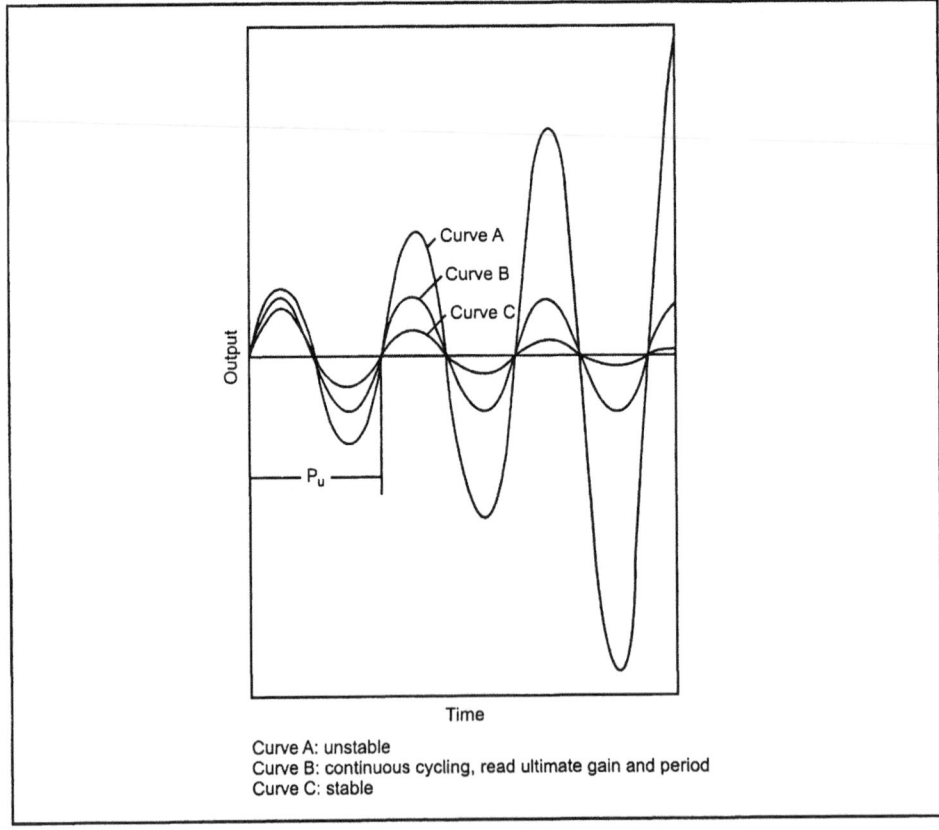

Curve A: unstable
Curve B: continuous cycling, read ultimate gain and period
Curve C: stable

Figure 9-3. Responses to Illustrate Ultimate Gain and Ultimate Period

To determine the ultimate gain and the ultimate period, perform the following steps:

1. Tune out all the reset and derivative action from the controller, leaving only the proportional mode. This implies that you set T_i equal to infinity and T_d equal to zero (or as close to these values as is possible on the controller).

2. Maintain the controller on automatic; that is, leave the loop closed.

3. With the gain of the proportional mode of the controller at some low value, impose an upset on the process and observe the response. One easy method for imposing the upset is to move the set point for a few seconds and then return it to its original value.

4. If the response curve from Step 3 does not damp out (as in Curve A in Fig. 9-3), the gain is too high (the proportional band setting is too low). The gain should be decreased (the proportional band setting should be increased) and Step 3 repeated.

5. If the response curve in Step 3 damps out (as in Curve C in Fig. 9-3), the gain is too low (the proportional band is too high), the gain should be increased (the proportional band setting should be decreased) and Step 3 repeated.

6. When you obtain a response curve similar to that of Curve B in Fig. 9-3, the values of the ultimate gain (or ultimate proportional band) setting and the ultimate period of the associated response curve are noted by the person doing the tuning. This ultimate gain where the sustained oscillations are encountered is the ultimate gain K_u and the ultimate period is P_u.

The ultimate gain and the ultimate period are then used to calculate controller settings. In the case of proportional controllers, Ziegler and Nichols correlated that a value of operating gain equal to one-half of the ultimate gain would often give a decay ratio of one-quarter, and they therefore proposed the following tuning rule of thumb for a proportional controller:

$$K_c = 0.5 \, K_u \qquad\qquad (9\text{-}1)$$

By similar reasoning and testing, the following equations were found to represent good rules of thumb for controller settings for more complex controllers. Proportional-plus-reset:

$$K_c = 0.45 \, K_u \qquad (9\text{-}2)$$

$$T_i = P_u/1.2 \qquad (9\text{-}3)$$

Proportional-plus-derivative:

$$K_c = 0.6 \, K_u \qquad (9\text{-}4)$$

$$T_d = P_u/8 \qquad (9\text{-}5)$$

Proportional-plus-reset-plus-derivative:

$$K_c = 0.6 \, K_u \qquad (9\text{-}6)$$

$$T_i = 0.5 \, P_u \qquad (9\text{-}7)$$

$$T_d = P_u/8 \qquad (9\text{-}8)$$

To repeat, you should remember that the preceding equations are empirical, so exceptions are inherent. They are intended to achieve a decay ratio of one-quarter, which is the inherent definition of good control.

Modern controllers, which are the type the student is likely to encounter, have gain dials or gain-adjustment provisions that are, as gain should be, dimensionless. The gain setting typically represents the ratio of milliampere (mA) output to the change in milliampere (mA) input to the controller. Controllers with sensitivity dials have gone out of use because of their lack of universality, but Exercises 9-3 and 9-4 at the end of this unit are based on sensitivity and are included in this book simply to illustrate the difference between sensitivity and gain.

It should be pointed out that the closed-loop ultimate method presented in this section is not very practical for slow processes, such as large kilns, fractionating units, or boilers. This is due to the fact that (a) it can take days for a pattern to develop, and (b) it would not be safe or economical to have the process oscillate.

Example 9-1: For a temperature control system whose ultimate gain K_u is 0.4 and ultimate period is two minutes, determine settings for various controllers using the Ziegler-Nichols ultimate method. Proportional:

$$K_c = 0.5 \, K_u = 0.2$$

Proportional-plus-reset:

$$K_c = 0.45 \, K_u = 0.18$$

$$T_i = P_u/1.2 = 1.67 \text{ min}$$

Proportional-plus-reset-plus-derivative:

$$K_c = 0.6 \, K_u = 0.24$$

$$T_i = 0.5 \, P_u = 1.0 \text{ min}$$

$$T_d = P_u/8 = 0.25 \text{ min}$$

9-4.   The Process Reaction Curve

One of the problems inherent in the closed-loop ultimate method is that you must cause the loop to oscillate to determine the ultimate period and the ultimate gain. Some control room operators have a real problem with this! On the other hand, open-loop methods usually require only that a single upset be imposed on the process. These methods provide more precise data about the dynamics of a feedback control system, and often they give slightly better tuning results—though the variation in satisfaction from loop to loop can be quite significant.

Not only may process reaction methods be applied to all processes. You may also apply them to those processes that are still on the drawing board by making engineering calculations about the flow and energy transfer rates as well as the mass and energy storage capacities of the processing unit to be controlled.

The process reaction curve is the reaction of the process to a step change in its input signal. The process reaction curve is the reaction of the "all else" shown in Fig. 9-2. It is important to have a complete picture of exactly what this process reaction curve might represent in a feedback loop, and this is shown in some detail in Fig. 9-4. Note that the air-to-valve signal can be used to introduce the step change to the overall process, and note also that the dedicated trend recorder on the feedback loop may be used to record the process reaction curve. In general, a process reaction curve can be determined as follows:

1.   Let the system come to steady state.

2.   Place the controller on manual operation, that is, remove it from automatic operation.

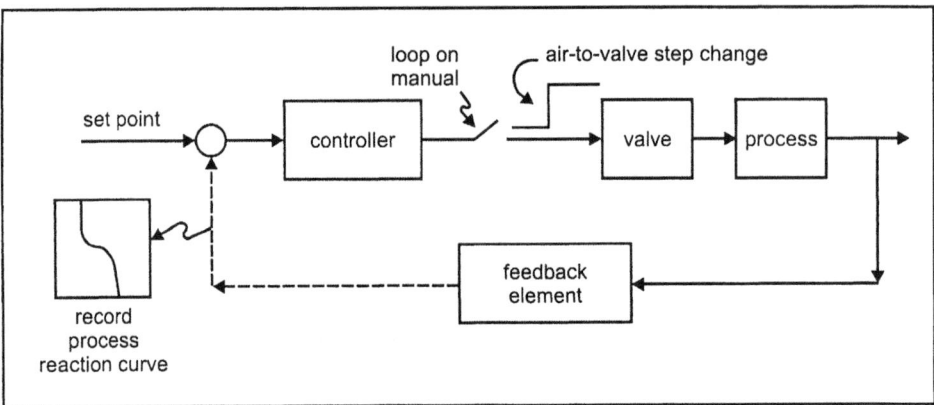

Figure 9-4. Determining the Process Reaction Curve

3. Manually set the air-to-valve signal at the value where it was operating automatically.

4. Allow the system to reach steady state.

5. With the controller still in manual operation, impose a step change in the air-to-valve signal. The size of the step change is significant because the process may exhibit nonlinear characteristics. Using a unit step change of the manipulated variable, typically 1 psi, is a common starting point.

6. Record the response of the controlled variable. (The response may be recorded by a dedicated trend recorder for the loop, but it is often desirable to have a supplementary recorder available that has a faster chart drive. This is because dedicated trend recorders have very slow chart drives and small charts.)

7. Return the air-to-valve signal to its previous value and return the controller to automatic operation. Once this has been done, the recorded process reaction curve may be used to give significant information about the overall dynamics and characterizing parameters for the "all else" of the process loop. This information may be used to calculate needed tuning parameters of the feedback controller.

9-5.   A Simple Open-Loop Method

One of the earliest methods that used the process reaction curve to tune controllers was also proposed by Ziegler and Nichols (in the same article that presented the ultimate method). To use this process reaction curve method, you must determine only the parameters R_r and L_r while knowing the sizes of the step change made and the sensitivity of the transmitter S_t which reports the process performance to the controller. Fig. 9-5 illustrates a sample determination of these parameters for a specific temperature control loop. This process reaction curve is based on a 1 psi step change.

To obtain process information parameters in the process reaction curve method, draw a tangent to the process reaction curve at its point of maximum slope. This slope is R_r which is the process reaction rate. Where this tangent line intersects the original baseline gives an indication

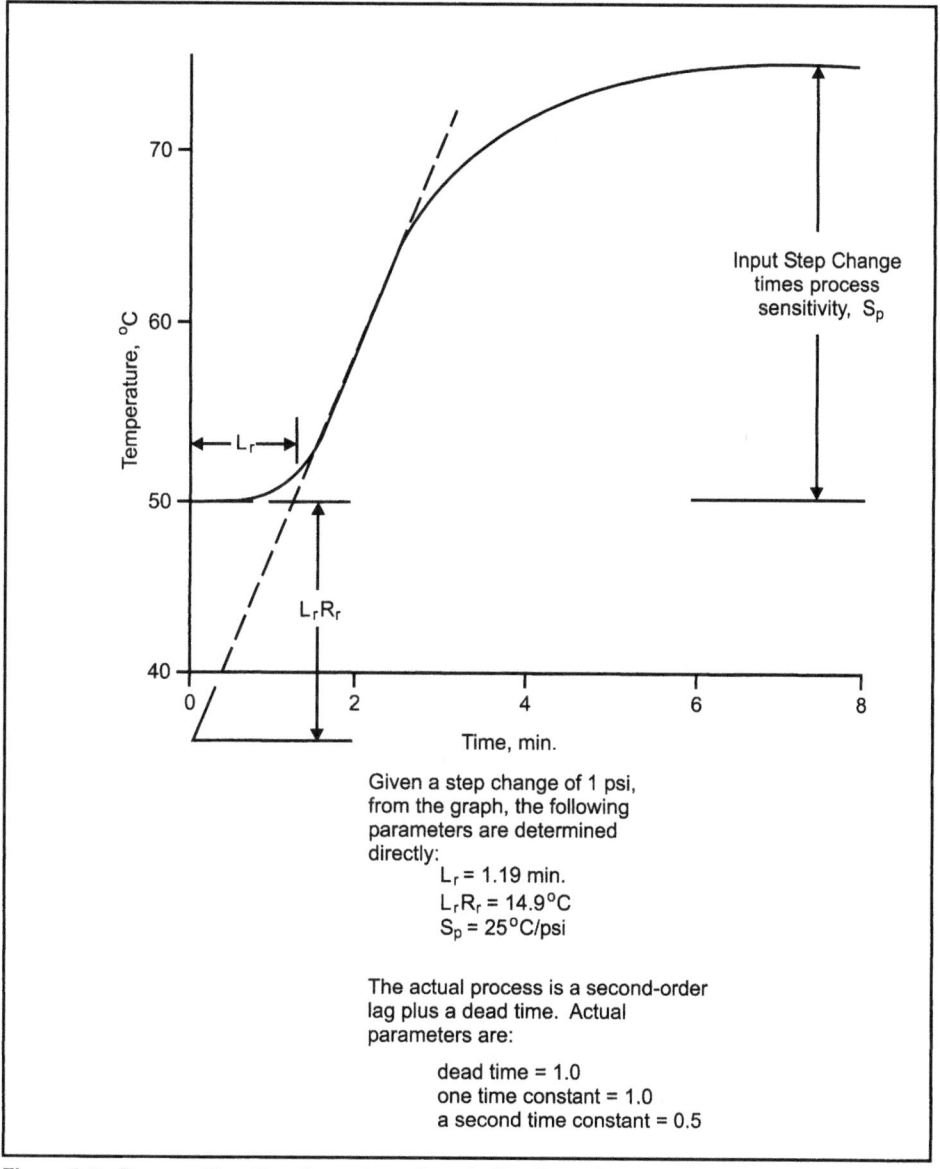

Given a step change of 1 psi, from the graph, the following parameters are determined directly:

L_r = 1.19 min.
L_rR_r = 14.9°C
S_p = 25°C/psi

The actual process is a second-order lag plus a dead time. Actual parameters are:

dead time = 1.0
one time constant = 1.0
a second time constant = 0.5

Figure 9-5. Process Reaction Curve for a Sample Feedback Control Loop

of L_r the process lag. L_r is really a measure of equivalent dead time for the process. If this tangent drawn at the point of maximum slope is extrapolated to a vertical axis drawn at the time when the step was imposed, then the amount that this tangent is below the horizontal base line will represent the product L_rR_r. Using these parameters, Ziegler and Nichols proposed a series of rules of thumb that can be used to calculate appropriate controller settings. Before this is done, however, you must determine the unit reaction rate, R_1:

$$R_1 = R_r\, S_t\, / \text{(Step size)}$$

Proportional only:

$$K_c = 1\, /\, L_rR_1 \tag{9-9}$$

Proportional-plus-reset:

$$K_c = 0.9\, /\, L_rR_1 \tag{9-10}$$

$$T_i = 3.33\, L_r \tag{9-11}$$

Proportional-plus-reset-plus-rate:

$$K_c = 1.2\, /\, L_rR_1 \tag{9-12}$$

$$T_i = 2.0\, L_r \tag{9-13}$$

$$T_d = 0.5\, L_r \tag{9-14}$$

Ziegler and Nichols indicated that these rules of thumb should give a decay ratio of one-quarter, which is their inherent definition of good control.

In summary, to use this particular process reaction curve method, it is necessary to obtain a specific process reaction curve with a known input step size and transmitter sensitivity. Then, using the L_r and the R_r that you graphically determined from the process reaction curve, calculate the needed tuning parameters.

Example 9-2: Fig. 9-5 shows a process reaction curve where it is determined that

Step Size = 1 psi

Transmitter sensitivity, $S_t = 12\ \text{psi}/300°C$

$$= 0.04\ \text{psi}/°C$$

$$L_r = 1.19 \text{ min}$$

$$L_r R_r = 14.9°C$$

Process Sensitivity, $S_p = 25°C/psi$

Therefore, $L_r R_1 = L_r R_r * S_t / \text{Step size}$

$$= 14.9 °C * (0.04 \text{ psi} /°C) /1 \text{ psi}$$
$$= 0.6 \text{ (a number without units, that is, dimensionless)}$$

Using Ziegler and Nichols's open-loop method, calculate the required controller settings:

Proportional only:

$$K_c = 1 / L_r R_1 = 1 / 0.6 = 1.7 \text{ (dimensionless controller gain setting)}$$

Proportional-plus-reset:

$$K_c = 0.9 / L_r R_1 = 0.9 / 0.6 = 1.5$$

$$T_i = 3.33 L_r = (3.33) (1.19 \text{ min}) = 3.9 \text{ min}$$

Proportional-plus-reset-plus-rate:

$$K_c = 1.2 / L_r R_1 = 1.2 / 0.6 = 2.0$$

$$T_i = 2.0 L_r = (2.0) * (1.19 \text{ min}) = 2.38 \text{ min}$$

$$T_d = 0.5 L_r = (0.5) * (1.19 \text{ min}) = 0.59 \text{ min}$$

9-6.   Integral Methods

In addition to the types of process tuning methods we have already presented, a large number of tuning techniques have been developed in recent years that are based on minimizing the values of various integral criteria. Usually, these techniques try to minimize the value of an error function over all time.[2] Integral criteria techniques are well suited for computer control applications and, in fact, are only recommended for such installations. In such cases, they can give very good tuning results.

---

2.   Common criteria minimize the error squared, the absolute value of the error, or the absolute value of error times time.

9-7.   The Need to Retune and Adaptive Tuning

Feedback control loops may be mathematically classified as *linear* or *nonlinear*. If a loop is linear it will simply double its output—via the exact-shape dynamic path—if you double its input. Similarly, if an input change is a mirror image of a previous input change, the output or response will be a mirror image. If a control loop is linear, it may be tuned once, and it will stay tuned forever! But if a loop is nonlinear, its tuning will require continuous adjustment (or it will be poorly tuned). This phenomenon is a major obstacle to keeping control systems well tuned.

It does not matter whether any one individual component of the control loop is linear or not. What matters is whether or not the "all else"—everything except the controller—is, in its aggregate form, linear or not. Sensors and transmission systems are normally designed by vendors to be linear. When one selects the valve trim, it is important to select valve trim and assign pressure drop across the valve so that its performance (see Fig. 7-8), when coupled with the performance of the process, produces a linear combination.

One very exciting aspect of the advances being made in the tuning of industrial control systems involves the concept of self-tuning or *adaptive tuning*. This idea is based on the concept that one should be able to design and implement controllers that are capable of tuning themselves or of being tuned automatically. Such control systems and individual controllers do exist, but until recently they were so expensive that they were not practical for routine application. The design and installation of systems and controllers capable of adaptively tuning themselves, that is, adapting themselves to the process situation that is manifest at a given moment, has been practical only for such situations as aircraft, special reactors, and paper machines. But microprocessors and digital capabilities have changed all this! Today, the computational capabilities required to do adaptive tuning are relatively cheap, and there is a strong movement toward the implementation of such systems.

The natural tendency when designing a self-tuning controller is to design an expert system that automatically performs the steps that an experienced control engineer would follow. This approach is basically one of "pattern recognition" in the closed-loop response of a control loop. To say this differently, the controller inspects the response pattern, measures overshoot and damping, and notes the times of successive peaks (thus establishing the period of oscillation). Based upon these measured parameters, you can tune the controller using conventional tuning approaches.

All of this sounds simple, but it is not. One of the most popular--and the first commercially successful--such single-loop, adaptive-tuning systems for the fluid process industries was a rule-based expert system. It had more than two hundred rules, most of which were to keep the pattern-recognition algorithm from being confused by peaks that were not caused by controller-tuning parameters.

An alternative approach to adaptive tuning is to use linear recursive regression to estimate the parameters of a discrete linear model of the process from the sampled values of the controller output and the controlled variable. Once the discrete process parameters have been calculated, they can be used to calculate controller parameters. Several commercially available controllers use this approach for adaptive tuning.

9-8. Summary

There are literally scores of different tuning techniques available, and they vary considerably in terms of philosophy and implementation. They also vary in terms of how they define good control, whether they are open-loop or closed-loop, how mathematically complex they are, and on and on. Quite often, they give widely varying results. It is impossible to say that one technique is clearly superior to all others.

It is fair to say that much of the tuning of control systems is an art. It requires a great deal of practice and experience to develop a sensitive and professional feel for the tuning of systems. Recognizing the many difficulties associated with tuning control systems, ISA has developed a specific book on the subject: *Tuning of Industrial Control Systems* by A. P. Corripio (ISA, 1990). The practitioner interested in developing more detailed insight into this particularly important subject should refer to this textbook.

REFERENCES

1.   J. G. Ziegler and N. B. Nichols, "Optimum Settings for Automatic Controllers," *Transactions ASME*, vol. 64 (Nov. 1942), p. 759. (This is a classic!)

Further Study

The tuning of control systems is such an important subject that ISA has published several specific books on the subject. These include the following:

1.   K. Astrom and T. Hagglund, *PID Controllers: Theory, Design, and Tuning*, 2d ed. (ISA, 1995).

2.   Armando B. Corripio, *Tuning of Industrial Control Systems* (ISA, 1990).

EXERCISES

9-1. *How is it physically possible in two-mode and three-mode controllers (such as PI or PID) for there to be more than one set of controller settings that give a one-quarter decay ratio?*

9-2. *Why don't all liquid-level control loops have the same tuning parameters, all flow control loops the same tuning parameters, and so on?*

9-3. *For a level control loop whose ultimate sensitivity is 0.3 psi/ft and whose ultimate period is three minutes, calculate the settings for the various types of controllers using the ultimate method. (Sensitivity is a gain with dimensions or engineering units; gain is dimensionless.)*

9-4. *Inspect Fig. 9-5 and note that the sensitivity S_p of the "all else" may be obtained. What are the units of this S_p? What are the units of the products of all the sensitivities around a feedback control loop, that is, the units of $S_p S_c$?*

9-5. *An experimental process reaction curve is run for a level control loop using a 1.2 psi step input. The level transmitter has a sensitivity of 12 psi/ 6 ft. From it, $L_r = 0.15$ min and $R_r = 0.6$ ft/psi/min. Calculate the controller settings for the various types of feedback controllers.*

9-6. *Convert the results of Exercise 9-5 for a controller using proportional band, reset rate, and derivative time.*

Unit 10:
Cascade Control

UNIT 10

Cascade Control

Until now, whenever we have mentioned feedback control in this book we have focused attention on simple, single-loop feedback control. In this unit, we will expand our horizons to include several variations of this basic single-loop approach.

Learning Objectives — When you have completed this unit, you should:

A. know the basic principles of cascade control

B. understand the basic concepts that govern the selection of the secondary controlled variable

C. understand the general guidelines for cascade controller mode selection and tuning

10-1. The Concept of Cascade Control

The general concept of cascade control is to nest one feedback loop inside another feedback loop. This is illustrated in Fig. 10-1. In effect, you take the process being controlled and find some intermediate variable within the process to use as the controlled variable for the inner loop.   The effect is to take the process and split or divide its dynamic lag into two parts.

Cascade control exhibits its real value when the dynamic lag is very large, that is, when you are attempting to control a very slow process. In a slow process, errors can exist for very long periods of time, and when disturbances enter the process there will be a significant wait before any corrective action is initiated. Also, when corrective action is taken, you will have to wait a long time for results. Cascade control affords you the

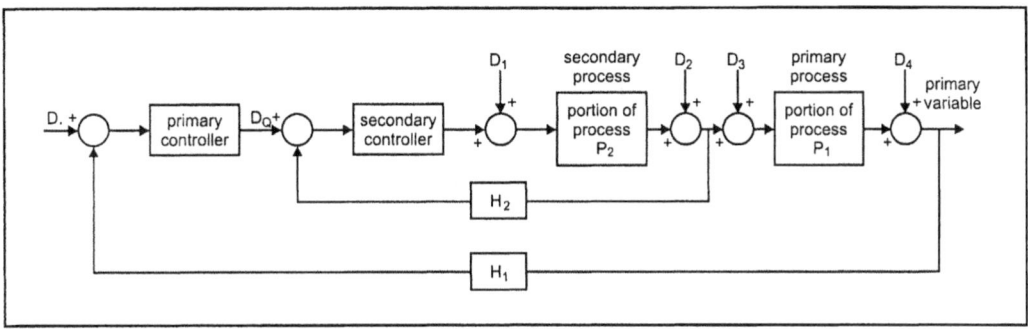

Figure 10-1. The Concept of Cascade Control

opportunity to find intermediate controlled variables and to take corrective action on disturbances more promptly.

Using cascade control would appear to involve significant additional hardware expenditures. As can be seen from the general layout in Fig. 10.1, it requires an additional feedback controller, and it appears to involve an additional sensor and feedback transmission system. There is no need, of course, for an additional final control element such as a control valve. All these general appearances are a bit misleading, however. Hardware vendors usually supply both the primary and secondary controllers for the cascade arrangement within a single controller case, and the total cost for the two controllers is not twice the cost of a single controller. In addition, it is probably true that the intermediate or secondary controlled variable is of sufficient importance that often it is already one of the variables that is sensed and transmitted back to the central control room for indicating and/or recording purposes. If this is the case, there is no incremental cost associated with the sensor and feedback transmission system of the secondary controlled variables. In the case of digital control, the extra need is for programming. In general, the incremental hardware and software costs associated with implementing cascade control are not prohibitive. As a matter of fact, these costs are not usually a significant factor.

In general, cascade control offers significant advantages to the user, and it is one of the most underutilized feedback control techniques. To put this differently, most plants could increase their usage of cascade control to significant advantage.

10-2. Simple Applications

Consider the room temperature control system shown in Fig. 10-2. For simplicity, assume that only heating is needed and that this is provided by the steam heat of a forced-air circulating system. In Fig. 10-2, a conventional thermometer measures room temperature and adjusts steam flow in a conventional feedback arrangement.

For the room heating system, assume that it is subject to several types of disturbances, such as variations in inlet air temperature and flow rate as well as variations in the heat load on the room. A reflective inspection of Fig. 10-2 leads to the conclusion that there is considerable time lag associated with the control of temperature in the room. The largest time lag is associated with the time it takes the room to change in temperature; this could reasonably be fifteen or twenty minutes.

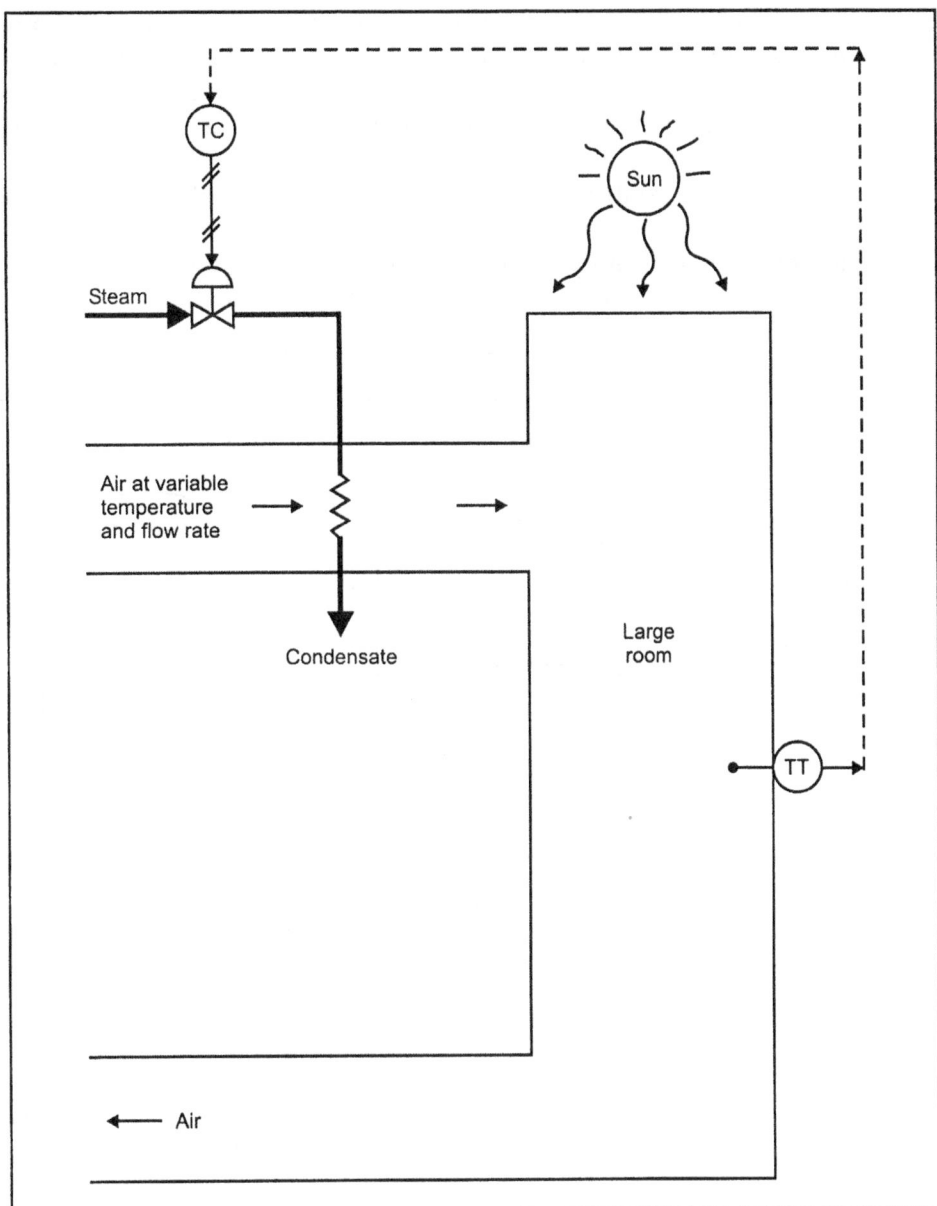

Figure 10-2. Room Temperature Control System I

There is also a lag in the corrective action associated with the change in temperature of the steam heater itself, for example, two or three minutes. For the sake of this example, ignore the time lags associated with the steam control valve and the room thermometer. Fig. 10-3 shows a cascade arrangement with a secondary temperature feedback control loop that measures and controls the temperature of the air entering the room. A primary temperature feedback control loop measures and controls room temperature by manipulating the set point or desired value on the secondary control loop for entering air temperature.

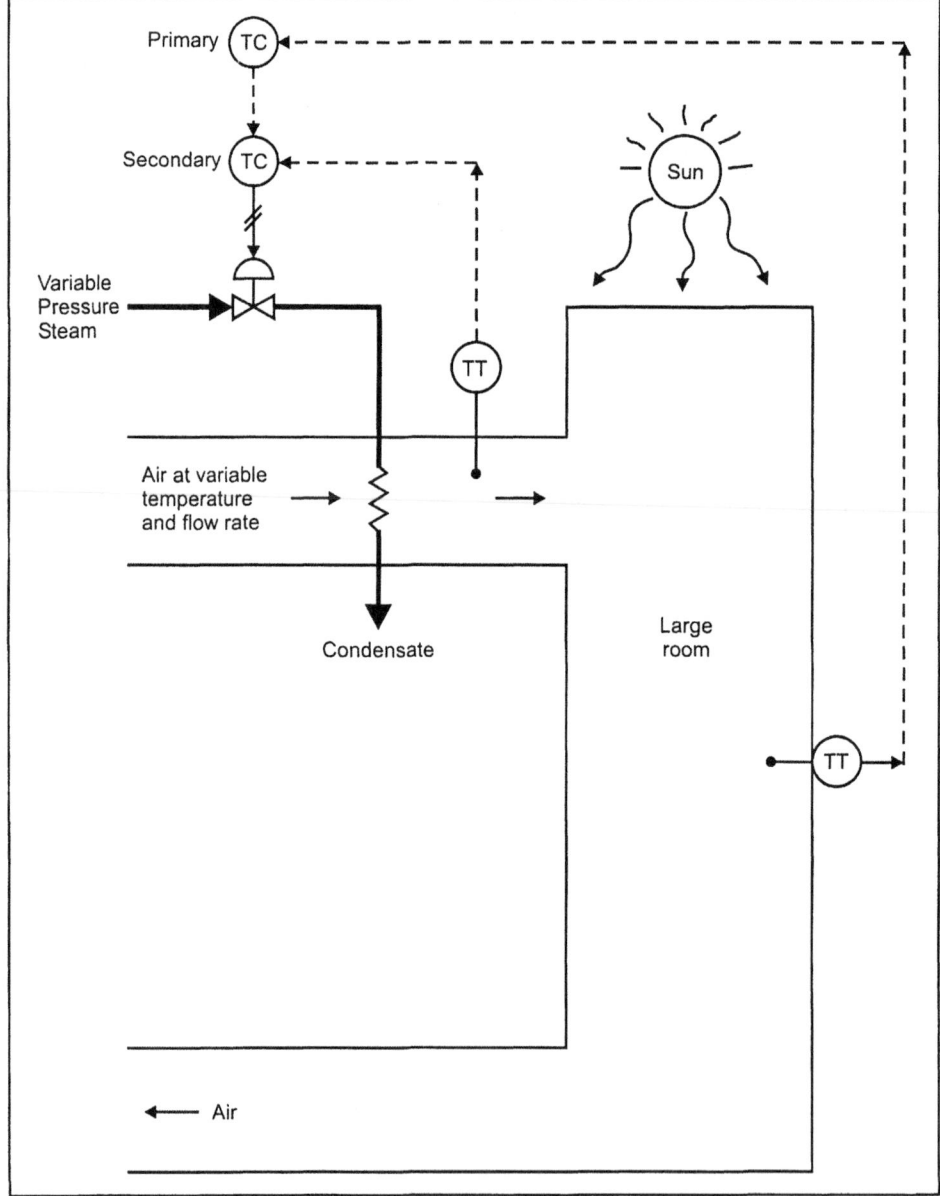

Figure 10-3. Room Temperature Control System II

With regard to the cascade system shown in Fig. 10-3, now consider how it will respond to a disturbance or variation in the entering air inlet temperature or flow rate. Clearly, it will be picked up by the secondary controller's sensor, which is located in the air duct, and corrective action can be taken immediately without waiting for deterioration in actual room temperature. A change in heat load within the room itself still must be picked up by the primary sensor that measures room temperature.

Fig. 10-4 shows a second simple example. In this case, several liquid streams flow together into a surge tank, and in turn they are fed into a distillation column. Fig. 10-4(a) shows a conventional feedback system with a level transmitter that senses and transmits tank level to a level controller that then adjusts a valve controlling feed into the column. For the sake of illustration, assume that the system is subject to disturbances caused by variation in inlet flows to the surge tank and also to disturbances caused by changes in column pressure. A change in column pressure will cause a change in pressure drop across the control valve and, thus, produce a variation in feed flow into the column.

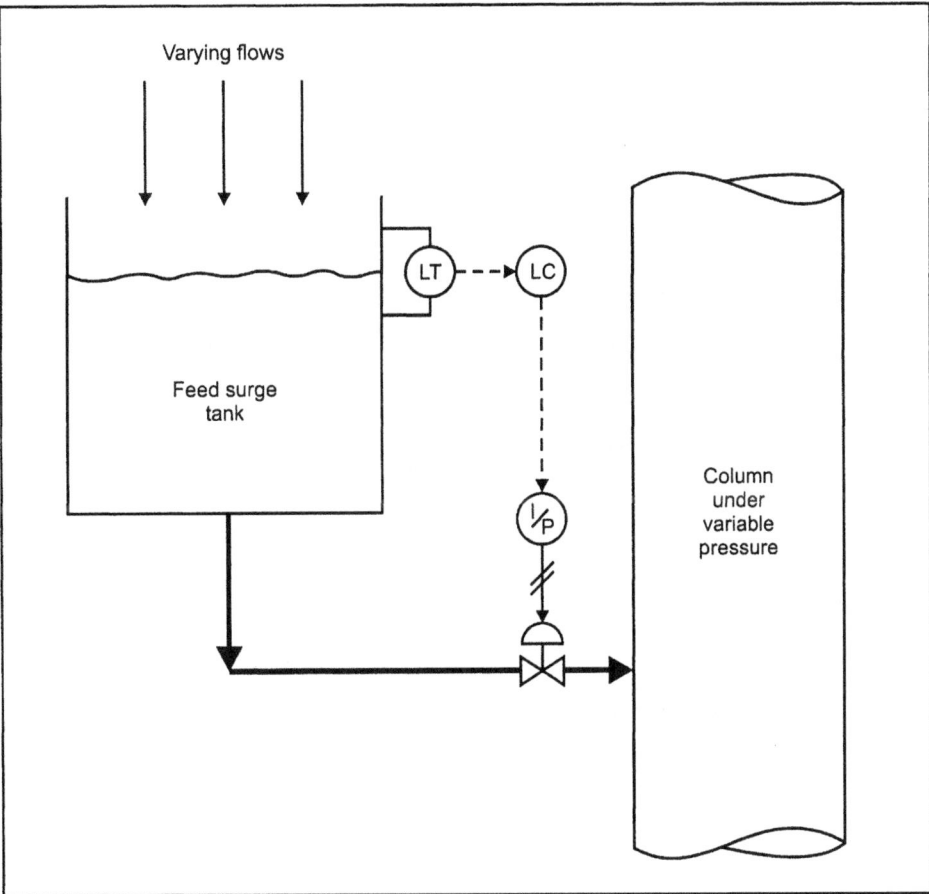

Figure 10-4(a). Column Feed Control System

Fig. 10-4(b) shows a cascade control system in which the feed into the column is flow controlled, and the set point on this flow control feedback loop is the manipulated variable of the primary level control feedback loop. In this case, when column pressure changes, pressure drop across the control valve and feed flow change, and this is sensed and controlled quickly by the flow control loop without having to await a resultant change in the surge tank level. In this case, it also is desirable to use a low sensitivity or gain in the controlled level so that changes in feed-flow set point are introduced slowly and the column is not subjected to rapid feed-flow rate changes. The usefulness and advantages of cascade control are obvious.

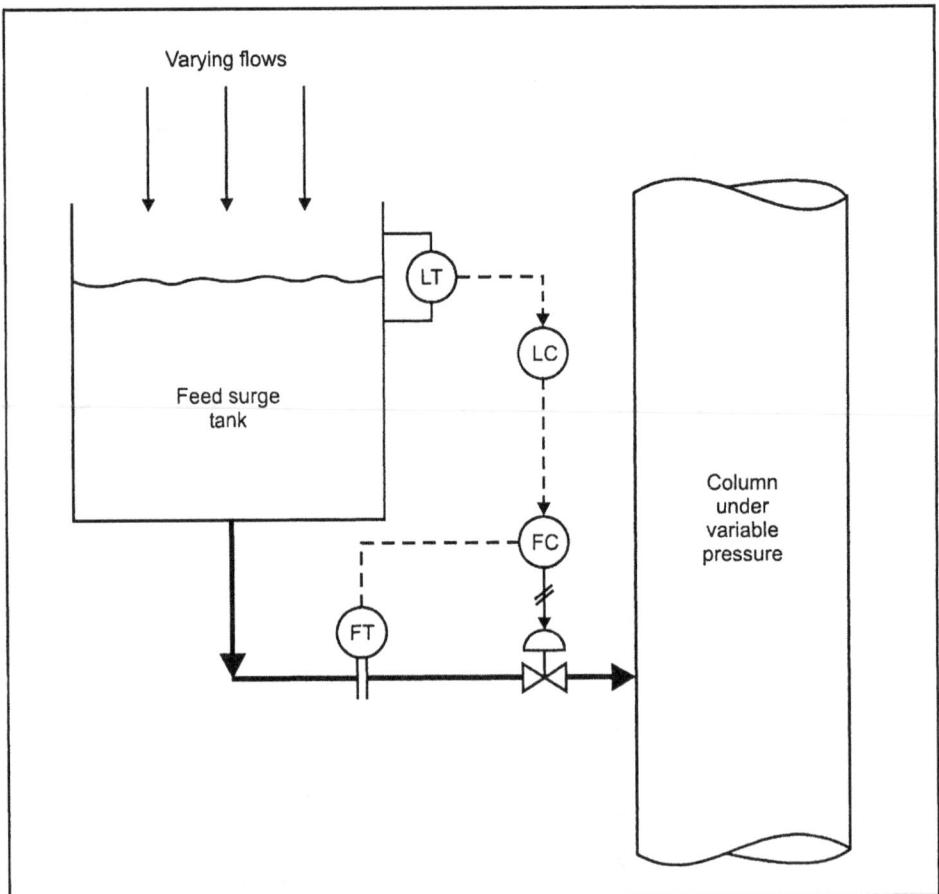

Figure 10-4(b). Column Feed Control System

10-3. More Complex Applications

Consider the catalyst regeneration system shown in Fig. 10-5(a), in which air is heated and then pumped through a regenerator where the hot air burns carbon off a catalyst. The exit gas temperature from the regenerator is controlled: too low a temperature requires excessive time; too high a temperature causes damage to the catalyst. A conventional temperature feedback system controls fuel to the heater. For simplicity's sake, Fig. 10-5(a) does not show air flow controls.

There are two major pieces of equipment within the feedback loop shown in Fig. 10-5(a), and it is clear that the lags involved can be split by measuring air temperature as it leaves the heater. The result makes it convenient to design the cascade control system shown in Fig. 10-5(b). The block diagram of this cascade system is shown in Fig. 10-6.

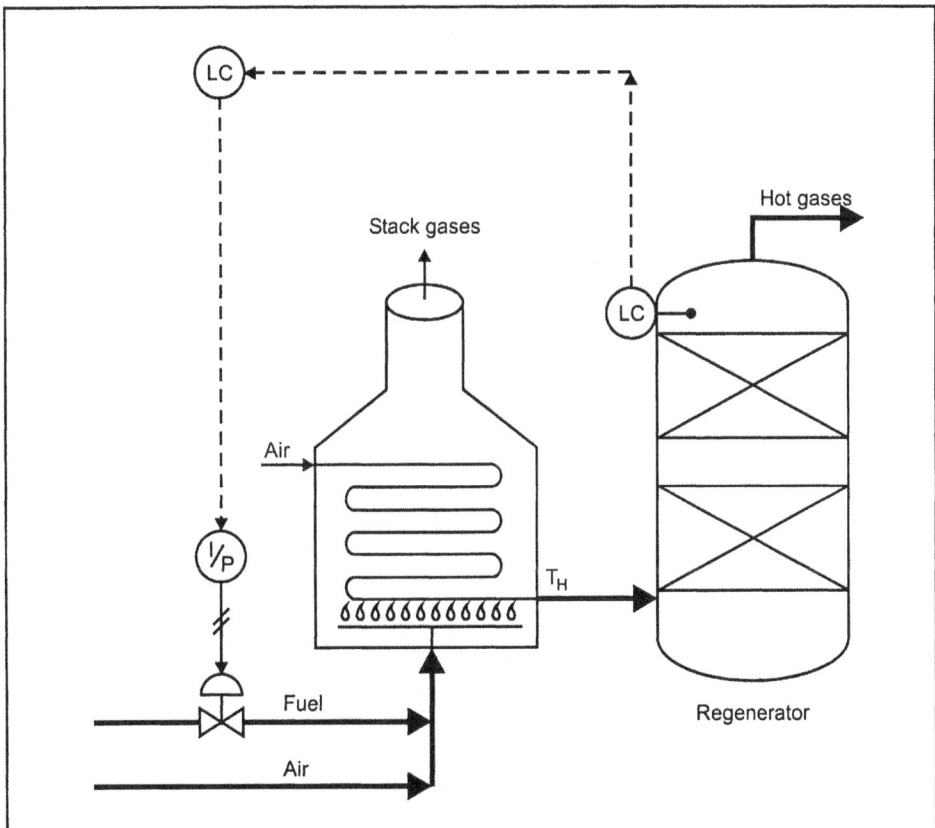

Figure 10-5(a). Catalyst Regeneration System

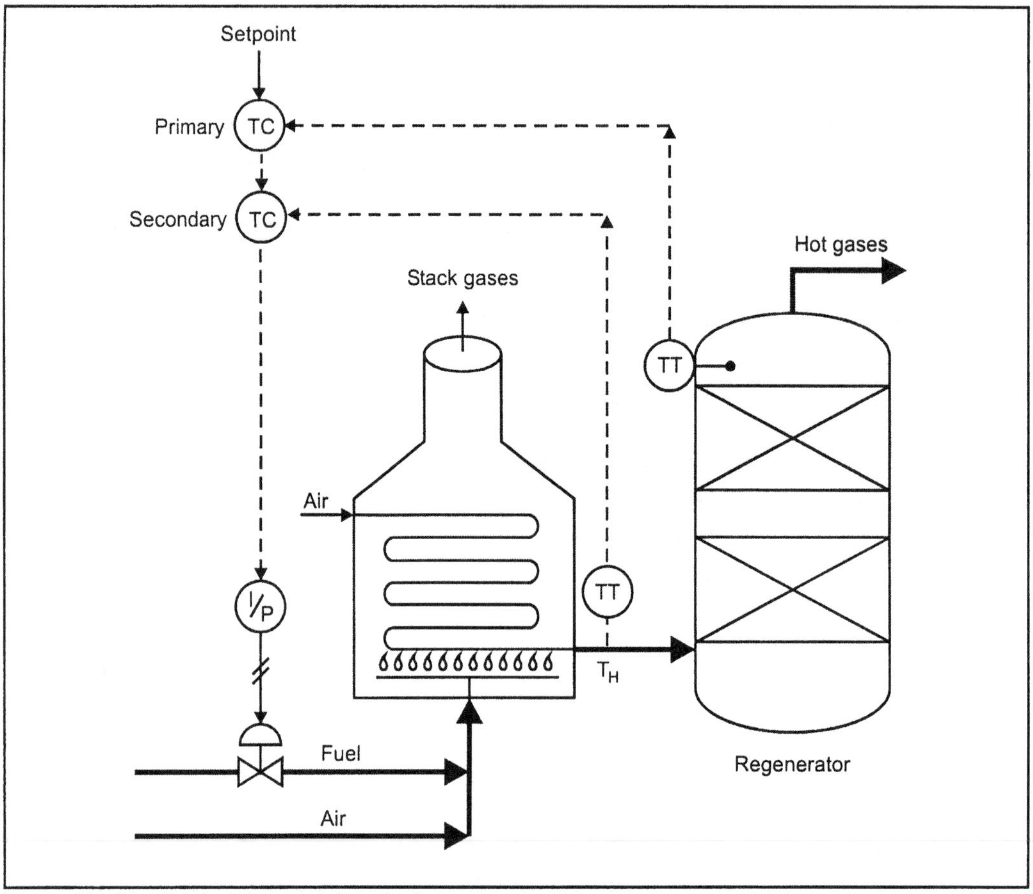

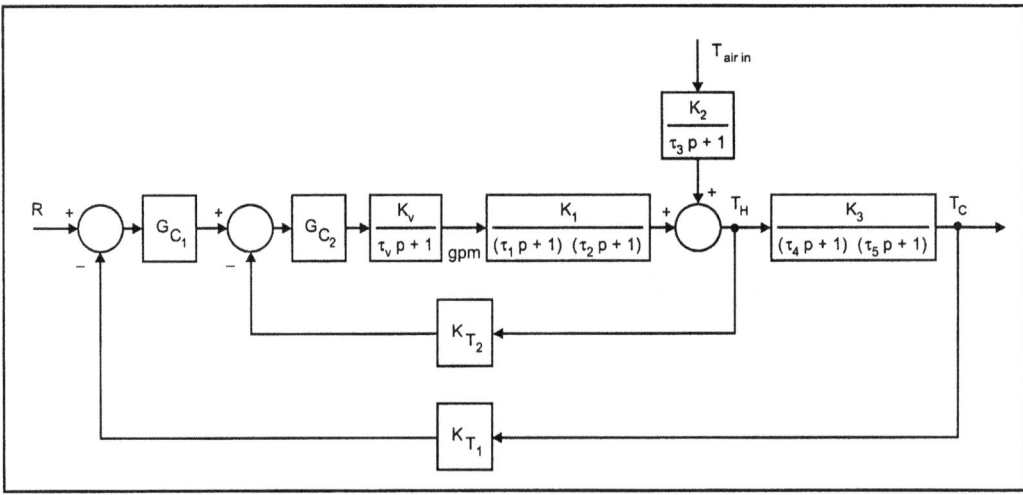

Figure 10-5(b). Catalyst Regeneration System

Figure 10-6. Block Diagram of Catalyst Regeneration System

Since in Fig. 10-5, two separate pieces of equipment exist, it is somewhat easy to split the process into two parts. However, in many cases, the time lags in a process are all present within a single piece of equipment. An example will further our understanding; consider Fig. 10-7. If this kettle had a simple feedback control loop without cascade provisions, the sensor on the primary controlled variable (the reacting mass temperature) would go to the feedback controller, and the output of the feedback controller would directly adjust the manipulated variable (the makeup cooling water supply). The jacketed kettle as presented in Fig. 10-7 has a cascade arrangement, with the temperature of the reacting mass remaining the primary controlled variable. An intermediate controlled variable for the secondary controller is the temperature of the circulating cooling water for the jacket of the kettle. In this case, as is standard in cascade control arrangements, the manipulated variable for the primary or master controller is the set point for the secondary or slave controller.

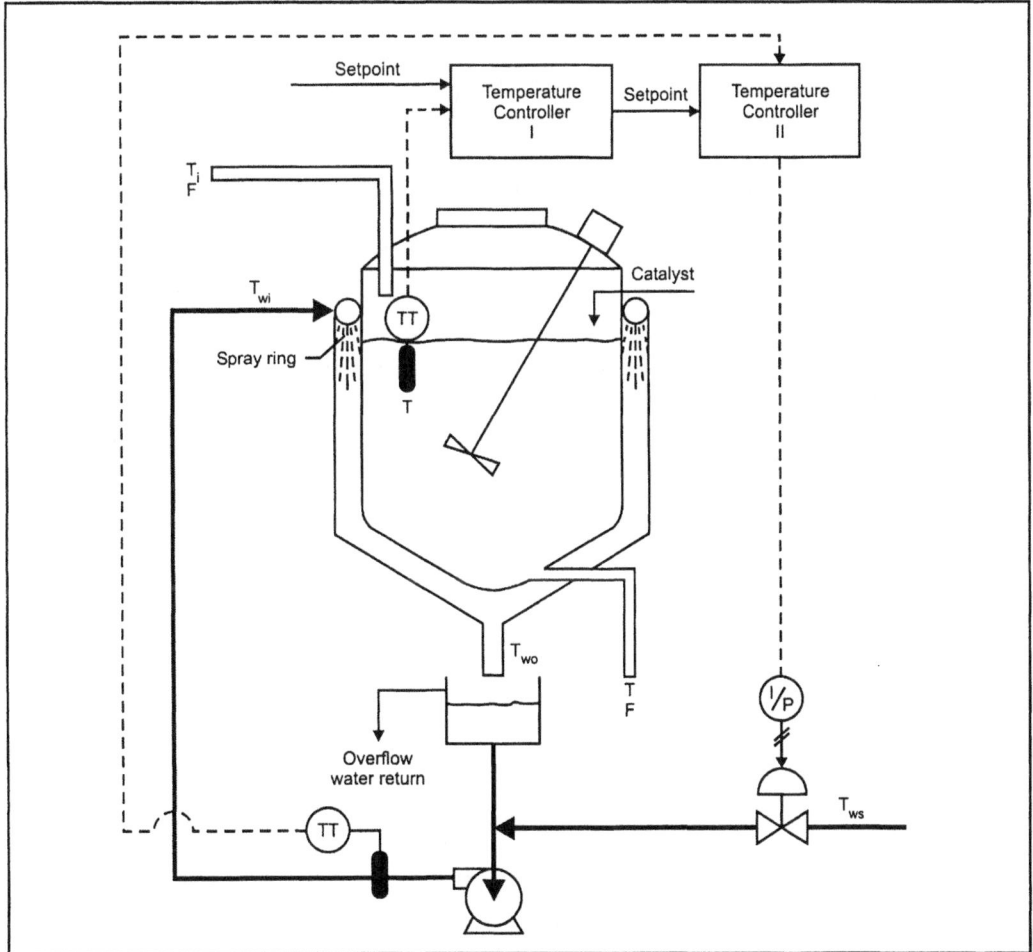

Figure 10-7. Cascade Control on a Jacketed Kettle

To understand cascade control, it will be helpful to look at two potential disturbances for the system in Fig. 10-7. First, consider what happens when there is a change in inlet feed temperature to the jacketed kettle. A change in this variable T_i will change the basic temperature of the reacting mass. In a general sense, the reacting mass will change in temperature by a first-order lag-type relationship. The time constant is the mass of the contents of the vessel times its heat capacity (capacitance) divided by the flow rate F of fluid into the vessel times its heat capacity (conductance).

This time constant for the contents of the kettle will probably be quite large. As the reacting mass starts to change in temperature, it will be sensed by the primary sensor, which has its own dynamics--at least a first-order lag--and this first-order lag is interacting with the time constant of the kettle contents. The primary controller receives the error caused by the change in temperature, and corrective action is taken by adjusting the set point on the secondary or slave controller. Then corrective action is taken by adjusting the manipulated variable, that is, the makeup cooling water flow.

As an alternative upset or disturbance, consider what happens if the cooling water supply has a change in its temperature T_{ws}. When this happens, there soon will be a change in the temperature of the discharge of the cooling water circulating pump, and this change in temperature will be sensed by the sensor for the secondary controller. If there were no cascade control, this disturbance would not be sensed until the temperature change worked through the entire cooling water system and began to change the temperature of the reacting mass inside the vessel. The advantages of the cascade arrangement for this type of upset are apparent.

Fig. 10-8 gives a general block diagram layout for this temperature control system, and the two disturbances are shown entering the system. The improvement in handling disturbances in the cooling water temperature is obvious.

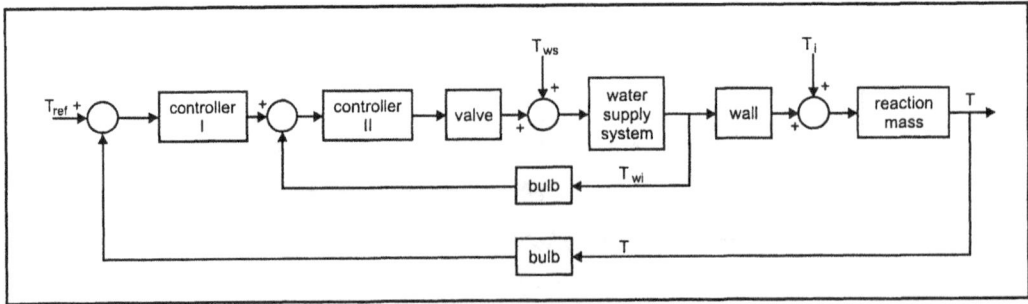

Figure 10-8. Block Diagram for Cascade Control on a Jacketed Kettle

10-4. Guiding Principles for Implementing Cascade Control

A tough question you may face when implementing cascade control is how to find the most advantageous secondary controlled variable, that is, determining how the process can best be divided. In the selection of this intermediate point, a large number of choices are quite often available to the designer. The overall strategy or goal should be to get as much of the process lag into the outer loop as possible while, at the same time, having as many of the disturbances as possible enter the inner loop.

Fig. 10-9 shows the general layout of a fired-charge heater that is used to increase the temperature of a fluid charge passing through a fired furnace. The feedback control of this arrangement is shown in Fig. 10-9(a). Fig. 10-9 also shows three different ways to establish a cascade control arrangement. In every case, the primary controlled variable is the same, but in each case a different intermediate controlled variable has been selected. Clearly, the question is: Which type of cascade control is best?

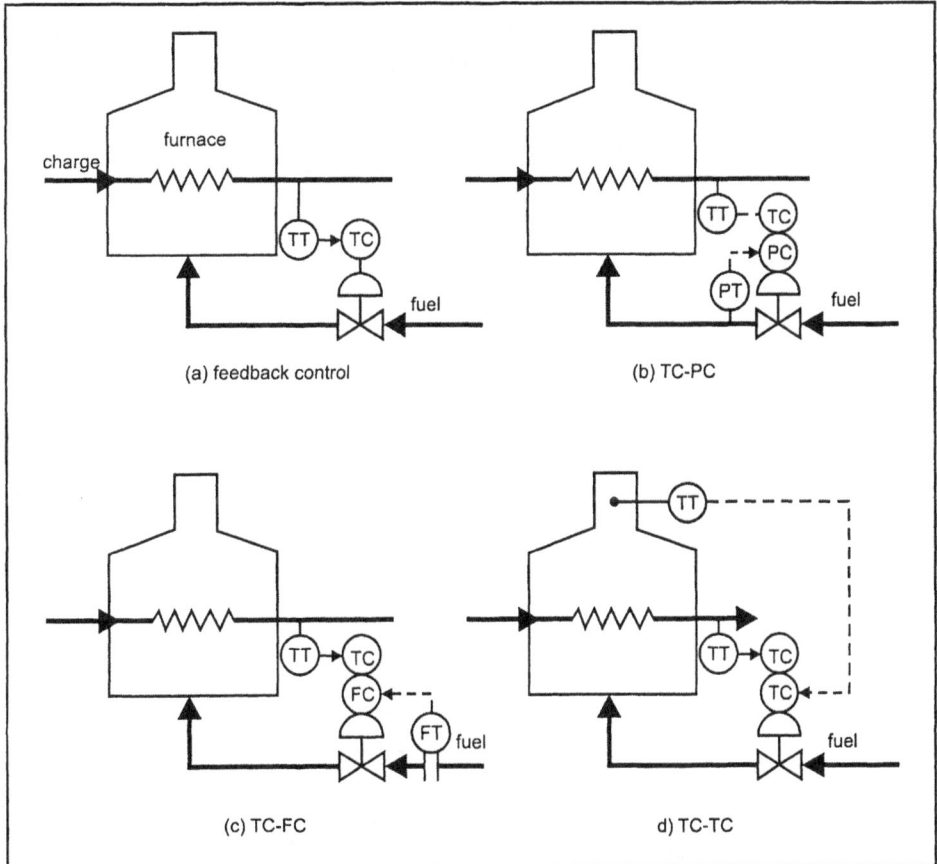

Figure 10-9. Alternate Cascade Control Arrangement on a Fired Charge Heater

To determine the best cascade control arrangement, you must make a specific determination of what the most likely disturbances to the system are. It is helpful to make a list of these in order of decreasing importance. Once you have done this, the designer may review the various cascade control options available and determine which one best meets the overall strategy outlined earlier, that is, to have the inner loop as fast as possible while, at the same time, receiving the bulk of the important disturbances. (Although it is almost of secondary interest to this particular discussion, in many petrochemical operations the basic argument is whether the TC-FC or the TC-TC arrangement is preferred. There is intense debate over which is the best arrangement.) As an example of the kind of thinking you should do, Fig. 10-10 shows the block diagram of the TC-FC fired-charge heater.

Selecting the secondary control variable is so important in a cascade system that you may find some formalized guidelines or rules of thumb will help guide the evaluation. The following rules reflect what we have learned so far:

Rule 1: Make the secondary loop include the input of the most serious disturbances. These disturbances that enter the secondary loop are the ones for which the cascade system will show the most improvement over conventional feedback control.

Rule 2: Make the secondary loop as fast as possible by including only minor lags of the complete control system. It is desirable–but not essential–that the inner loop be at least three times as fast as the outer loop.

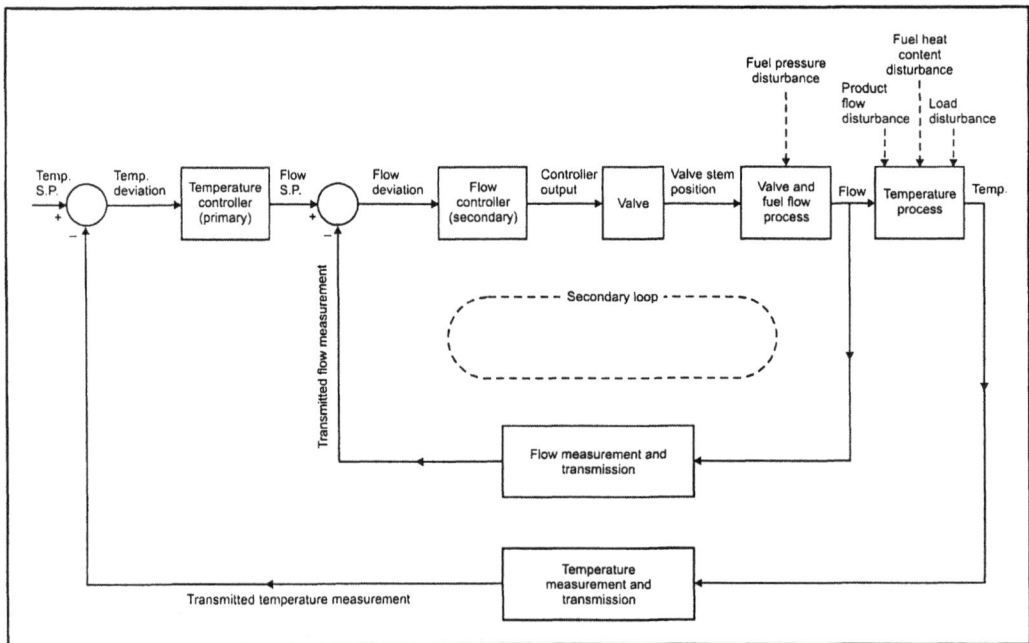

Figure 10-10. The Block Diagram of TC-FC Fired Charge Heater

Rule 3: Select a secondary variable whose values are definitely and usefully related to the values of the primary value. During undisturbed operation, the relationship between the secondary variable and the primary variable should be represented by a single line. If it is a straight line, the tuning of the controllers is much simpler.

Rule 4: If it can be done while keeping the secondary loop relatively fast, have the secondary loop contain as many of the disturbance inputs as possible.

Rule 5: Choose a secondary control variable that will allow the secondary controller to operate at the highest possible gain (lowest proportional band). This is difficult to predict.

To see how these rules are applied, consider the jacketed steam kettle shown in Fig. 10-11. The primary controlled variable is the temperature of the liquid in the vessel, and this is an inherently very slow-changing variable. Assume that steam flow, jacket pressure, and jacket temperature are all subject to major uncontrolled changes and that all could be the secondary loop's controlled variable. Table 10-1 gives the three possible secondary variables. The comparisons are made in light of the five rules of thumb just given. Fig. 10-12 illustrates the choices of secondary controlled variable.

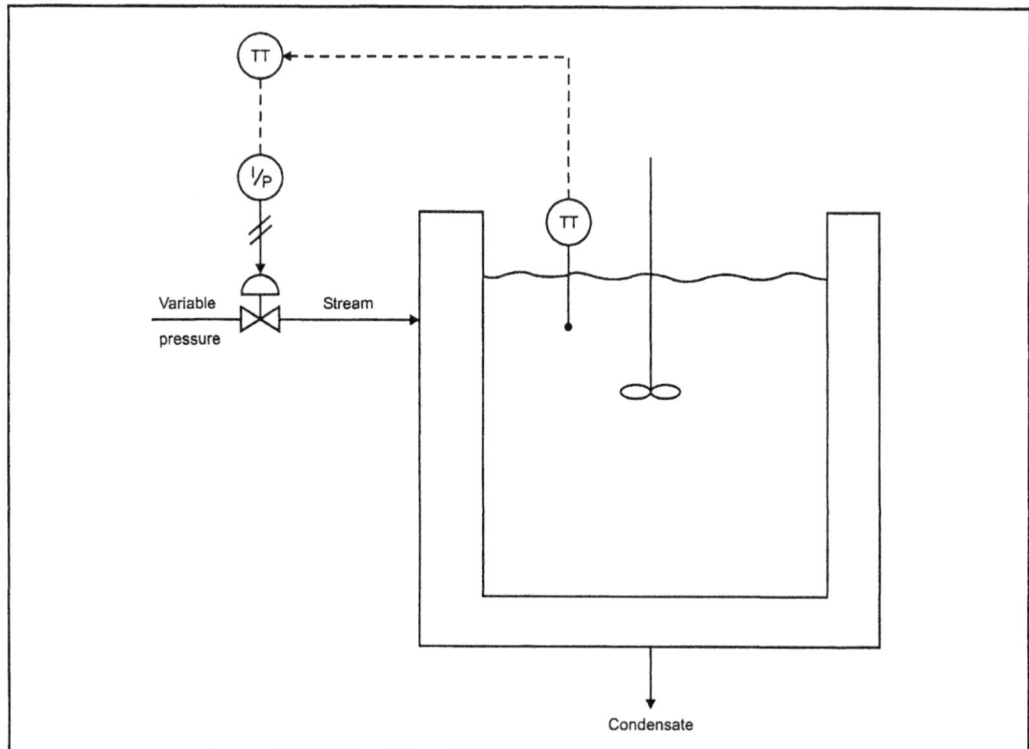

Figure 10-11. Temperature Control of Liquid in a Jacket Kettle

For this jacketed kettle, the TC-TC arrangement is best—and should be especially helpful as compared to conventional feedback control.

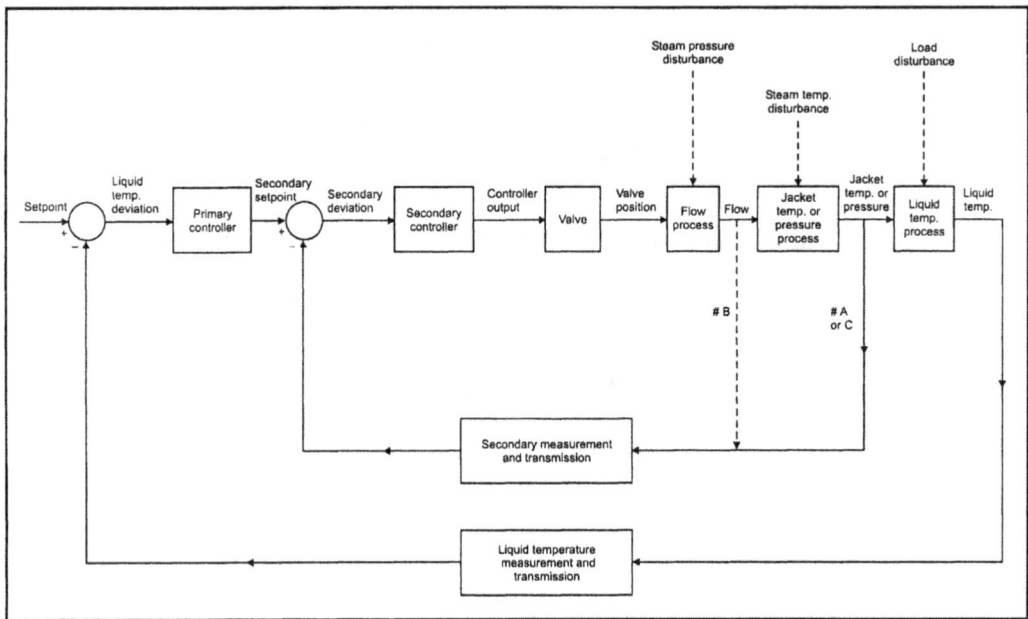

Figure 10-12. Block Diagram of Possible Cascade Arrangements

Condition Required by Rule of Thumb	Possible Secondary Variable		
	Jacket Pressure A	Steam Flow B	Jacket Temperature C
The most serious disturbances enter in secondary loop	Yes	Yes	Yes
Primary loop effective time constant is at least three times as long as secondary loop effective time constant	Yes	Yes	Yes - See Note 1
Values of secondary and primary are definitely and usefully related	Possible - see Note 2	See Note 3	Yes
Additional disturbances inputs are in secondary loop (See Note 4)	No	No	Yes
Proportional band of the secondary controller can be relatively narrow	Yes	No - see Note 5	Yes

Notes:

1. Jacket temperature will not change nearly as fast as either flow or jacket pressure, but it is still probably five or more times faster than the response of the liquid process mass inside the kettle itself.

2. Yes, since this is a condensing steam situation in the jacket, and pressure determines the condensing temperature.

3. The use of temperature to reset a flow set point can be a problem since flow loops usually have a square root scale. If temperature varies widely, tuning the controllers can be a problem and may necessitate the use of either a linear flowmeter or a device that varies the flow set point as a square of the primary controller output.

4. The "Yes" here is especially advantageous because it is obtained without slowing the secondary loop further.

5. This rule allows you to take advantage of the best possible reduction in deviations, but it will not normally be the deciding condition.

Table 10-1. Cascade Comparisons (Ref. 1)

10-5. More Examples

Cascade control as a concept is best understood through examples, and because of the importance of the concept, several more examples are presented here. Fig. 10-13 shows a TC-PC control system on a shell-and-tube heat exchanger. Pressure is used to indicate the tube wall temperature since the liquid film offers a much greater resistance to heat transfer. The tubes will, for all practical purposes, be at the condensing temperature that corresponds to the steam pressure in the shell.

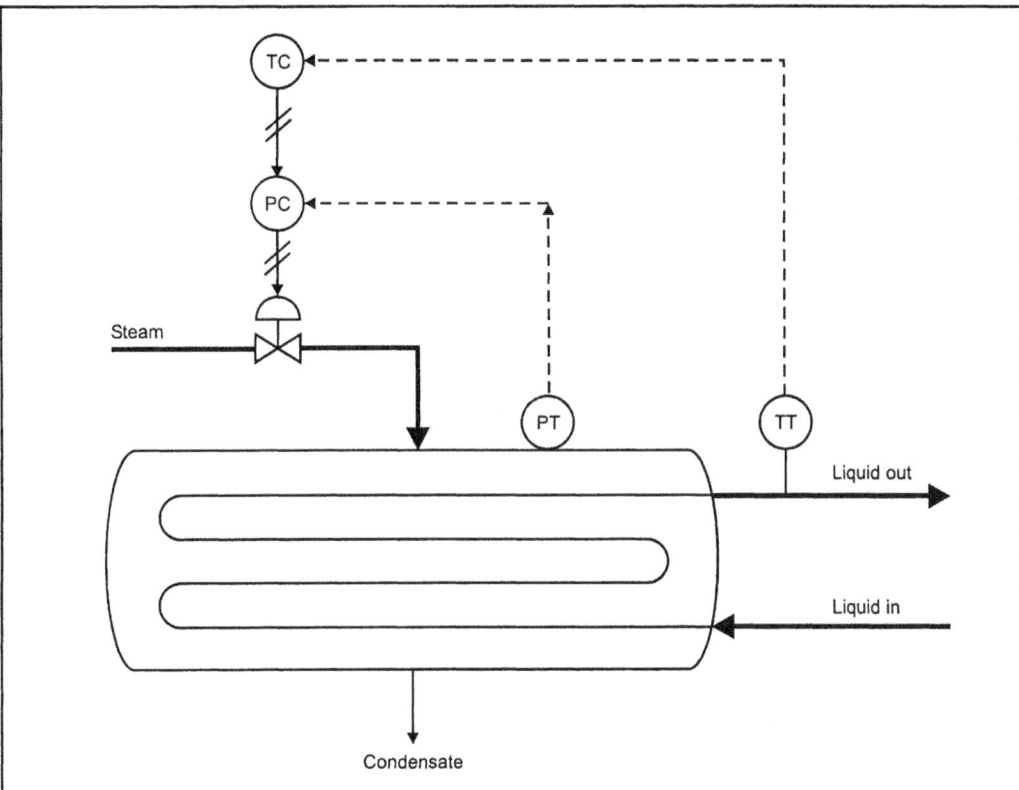

Figure 10-13. Shell-and-Tube Heat Exchanger Cascade Control

Fig. 10-14 shows a column being fed with a slurry that tends to build up in the control valve at small openings, but that clears when the valve opens sufficiently. Secondary loop control of flow handles this well.

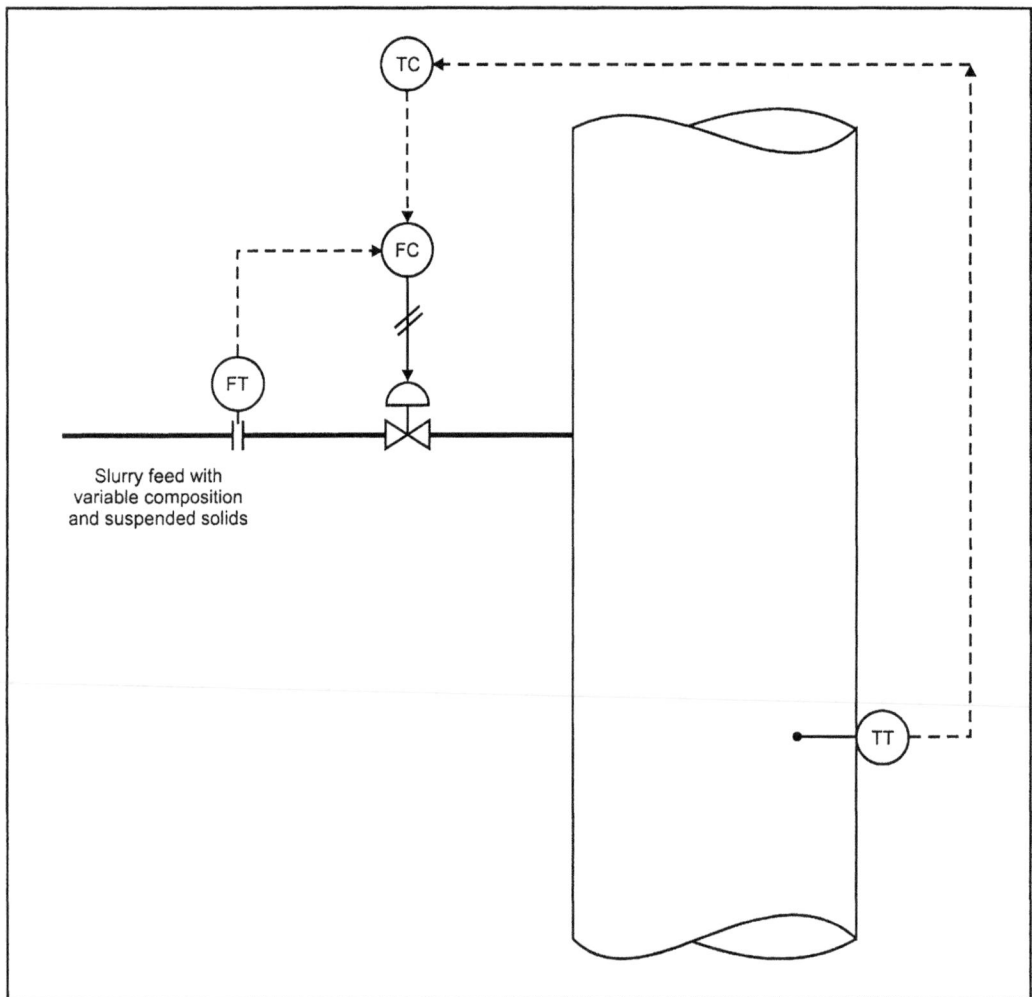

Figure 10-14. Slurry Feed to Column

Finally, Fig. 10-15 shows a situation in which the secondary controlled variable is heat-flow rate in a reboiler—a quantity that itself must be calculated from flow and temperature increase. Flashing occurs across the control valve.

The many examples we have presented in this unit serve to illustrate the value and potential of cascade control, which is underutilized in the process industries.

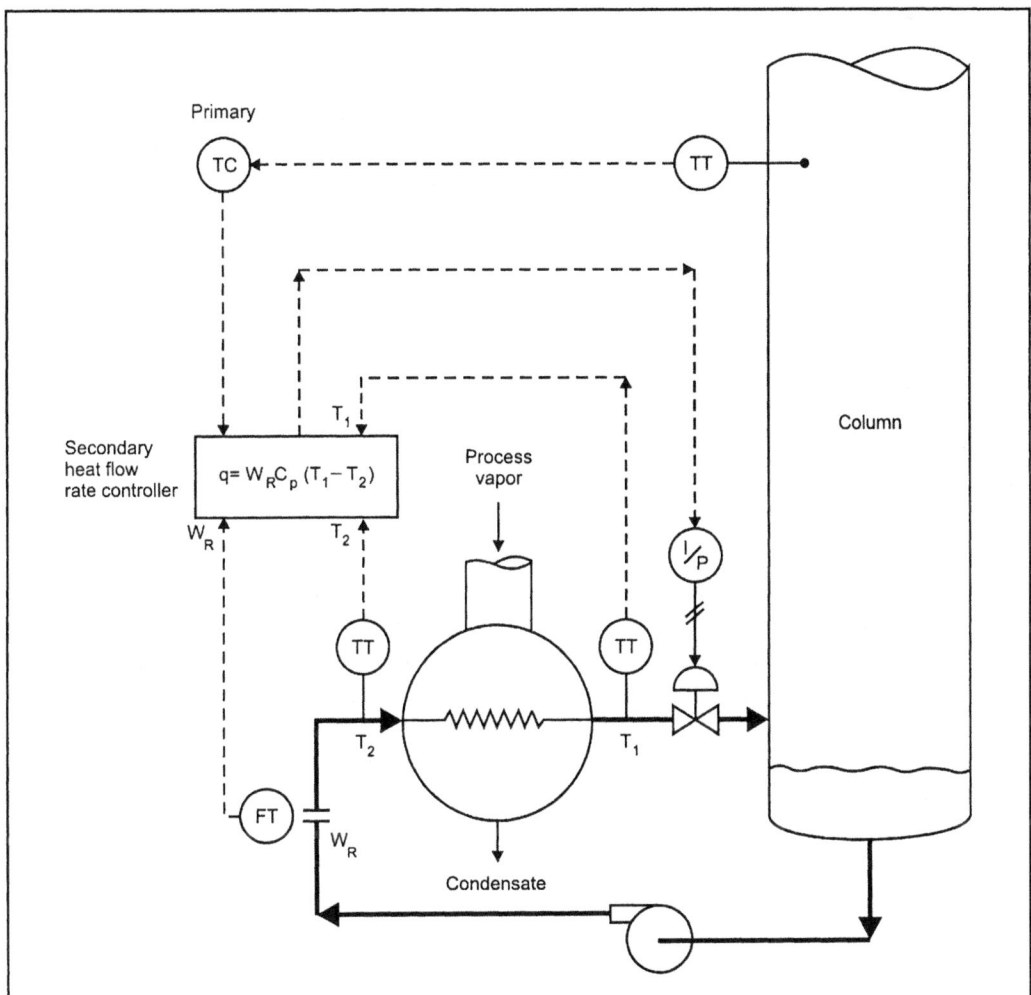

Figure 10-15. Cascade Control Using Heat Flow Rate in Secondary Controller

10-6. Selection of Cascade Controller Modes and Tuning

If both controllers of a cascade control system are three-mode controllers, there is a total of six tuning adjustments. It is doubtful if such a system could ever be tuned in an effective way.[1] When selecting the modes to be included in both the primary and secondary controllers of a cascade arrangement, the burden of proof should be on proving why a control mode should be added. For the secondary (or inner or slave controller), it is standard practice to include the proportional mode. There is little need to include the reset mode in order to eliminate offset since the set point for the inner controller will be reset continuously by the outer or master controller. Reset can be added to the inner-loop controller if it is a liquid-flow loop because of the need to filter some of the transmission of high-frequency noise around the loop. An alternative is to operate the inner loop at a low gain (high-proportional band). We discussed this earlier in Unit 9.

For the outer loop, the controller should contain the proportional mode, and if the loop is sufficiently important to merit cascade control it is probably true that reset should be included to eliminate offset in the outer loop. You should use rate or derivative control in either loop only if the loop has a very large amount of lag.

Tuning cascade controllers is the same as tuning all feedback controllers, but the practitioner must *work from the inside out*. The master controller should be put on manual, that is, the loop should be broken, and then the inner loop can be tuned by using whatever tuning technique the practitioner finds most useful. Once the inner loop is properly tuned, then the outer loop may be tuned. When it is, the outer loop "sees" the tuned inner loop functioning as part of the total process or as part of everything else that is being controlled by the master controller. If you follow this general inside-first principle when tuning cascade controllers, you should encounter no special problems.

REFERENCES

1.   D. M. Wills, *Cascade Control Applications and Hardware*, Technical Bulletin No. TX 119-1 (Philadelphia: Minneapolis-Honeywell Regulator Co.). (This is a classic!)

---

1.   Remember Murrill's Law: The difficulty of tuning a control system increases with the square of the number of modes present.

EXERCISES

10-1. *Given the following heat exchanger,*

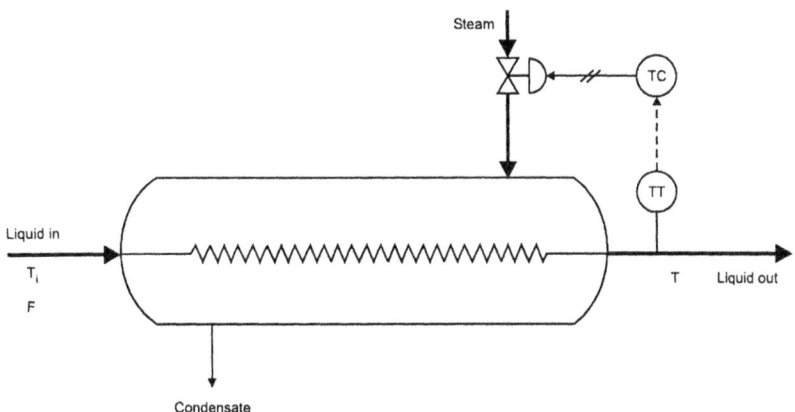

sketch a cascade arrangement for TC-FC control in which a secondary flow loop has been provided for the steam.

10-2. *Redo Exercise 10-1, but provide TC-PC control with a secondary pressure control loop on the exchanger shell.*

10-3. *Compare the results of Exercises 10-1 and 10-2. Which gives better initial response to disturbances in inlet process temperature T_i?*

10-4. *Given the following composition controller CC on the reflux flow to a distillation column,*

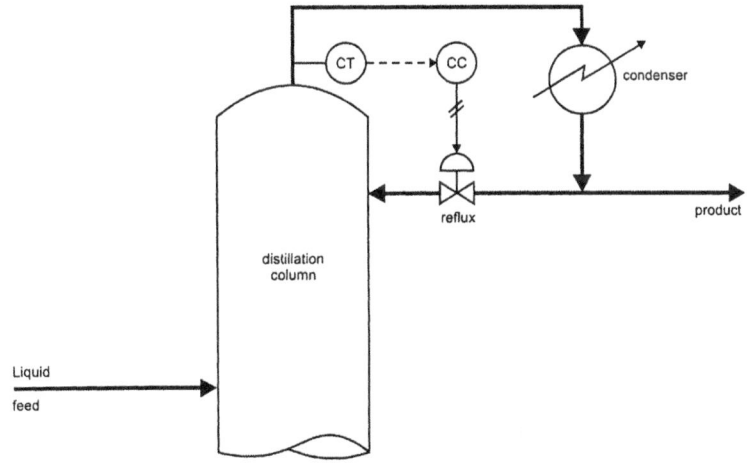

add a liquid-flow controller on the reflux as a cascade arrangement. Now add a level controller on the reflux condenser to govern product withdrawal.

10-5. *Given the following reboiler level control system,*

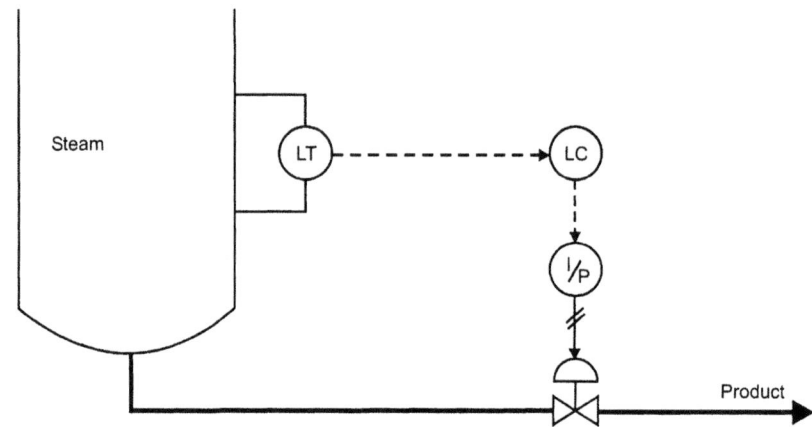

assume column pressure varies. Design a cascade arrangement to handle the product withdrawal.

10-6. *Given the following,*

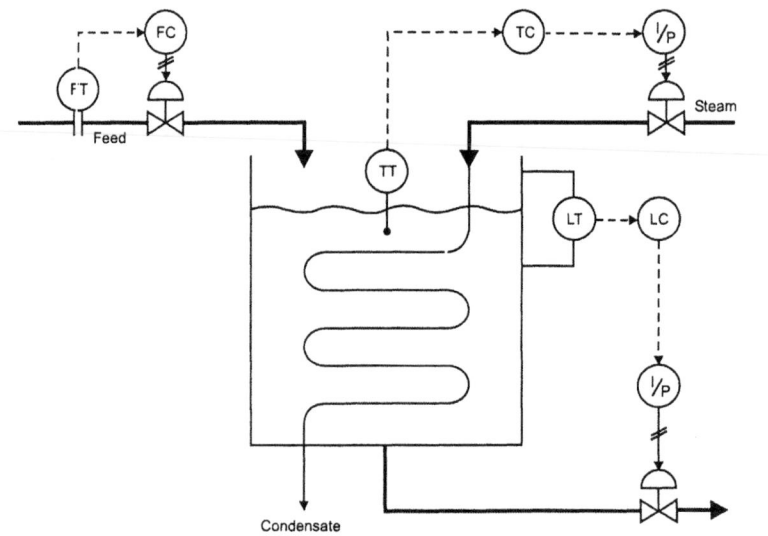

sketch a TC-PC cascade arrangement with steam pressure as a secondary control variable.

10-7. *Refer to the jacketed kettle of Fig. 10-7. Are you better off or worse off with cascade control (as compared to plain feedback control) if the inlet feed temperature T_i changes?*

10-8. *Given the cascade control system of Fig. 10-5(b), add flow control of the fuel to the heater. Sketch a diagram of this new control system.*

Unit 11:
Feedforward and
Multivariable Control

UNIT 11

Feedforward and Multivariable Control

All of the previous units have focused on feedback control. In this unit we will focus on some of the inadequacies of feedback control and show how feedforward control can be used to advantage.

Learning Objectives — When you have completed this unit, you should:

A. understand the basic concepts of feedforward control

B. be able to outline the structure of the control equation contained in a feedforward controller

C. understand the problems associated with multivariable control and the solution approaches available for solving these problems

11-1. Feedforward Control

There are two process conditions that can make the overall effectiveness of feedback control quite unsatisfactory. One of these is the occurrence of disturbances of large magnitude, and the other is the occurrence of large amounts of process lag. The question of how important either of these occurrences is is defined in economic terms. In either case, the principal concern is the existence of errors that have significant economic consequences on overall process operations. Feedforward control can be used to deal with these disadvantages or inadequacies of feedback control, that is, to deal with these errors.

We introduced the general concept of feedforward control in Unit 2. It is now time to view this concept in terms of a single loop. Fig. 11-1 shows the possibilities of a single controlled variable for a process subject to significant disturbance. A manipulated variable is chosen, and the selection of the manipulated variable is based on many of the same criteria typically used in feedback control.

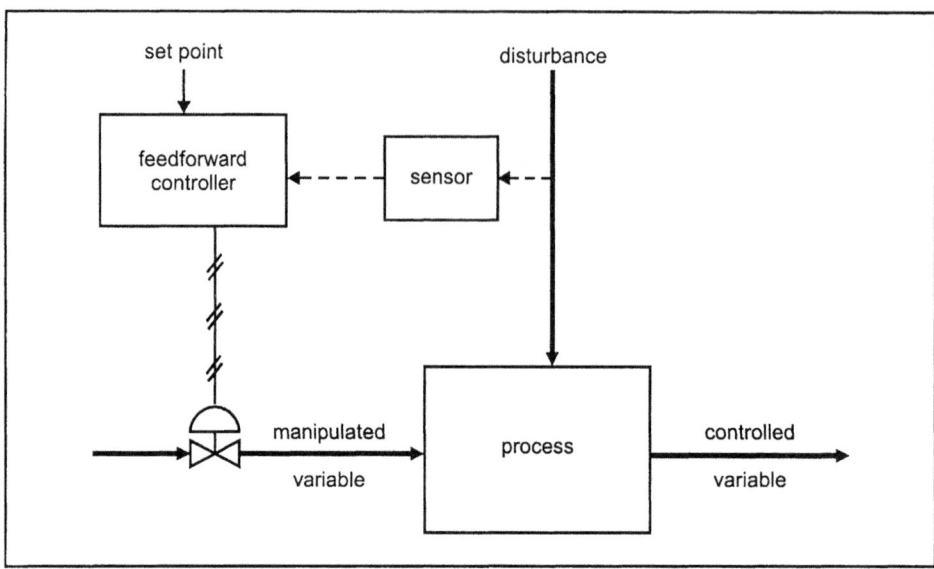

Figure 11-1. Feedforward Control Loop

In feedforward control, however, a sensor is used to measure the disturbance as it enters the process, and the sensor transmits this information to a feedforward controller. The feedforward controller determines the change that is required in the manipulated variable so that when the effect of the disturbance is combined with the effect of the change in the manipulated variable, there will be no change in the controlled variable. This perfect compensation is a difficult goal to obtain. It is, however, the objective that feedforward control is structured to achieve. Just as in feedback control, it is necessary to provide the feedforward controller with a set point or desired value for the controlled variable.

There are some significant difficulties involved in using feedforward control. The structure of feedforward control assumes that the disturbances are known in advance, that the disturbances will have sensors associated with them, and that there will be no significant unmeasured disturbances. A tremendous escalation of theoretical know-how in the feedforward controller's computation activities is also required. In feedback control, relatively standard control algorithms (such as the P, PI, or PID controllers) were used, but in feedforward control each controller equation or algorithm is specifically and uniquely designed for the one particular control application involved. In effect, the feedforward control computation involves determining exactly how much change in the manipulated variable is required for a specific change in disturbance. To be able to make this computation accurately requires significant quantitative understanding of the process and its operation.

There is one other significant aspect of pure feedforward control as shown in Fig. 11-1 that merits consideration. In this example, there are no feedback phenomena whatsoever; if the controlled variable strays from its set point or desired value, the control system is unaware of this and takes no corrective action to eliminate the deviation. This makes pure feedforward control somewhat impractical and a rarity in typical process applications. Therefore, the usual case is to combine feedforward control with feedback control.

It can be seen that feedforward control requires a significant increase in technical skills and capabilities. As a result, feedforward control of specific variables is limited to the most economically significant cases. In practical industrial applications, only very few cases are handled with feedforward control. These cases are, however, usually very significant, and the overall subject of feedforward control thus merits study.

11-2. A Feedforward Control Example

Fig. 11-2 shows a simple process heat exchanger. In this heat exchanger, a liquid flows through the exchanger and is heated by condensing steam. The controlled variable for this simple example is the exit temperature of the liquid flowing through the exchanger. The manipulated variable is the steam flow to the exchanger, and it has its own cascaded feedback control loop. Basically, the feedforward control is needed to calculate the desired value of this manipulated steam flow. The significant disturbances to this particular process are assumed to be the inlet liquid temperature and the flow rate of liquid through the exchanger.

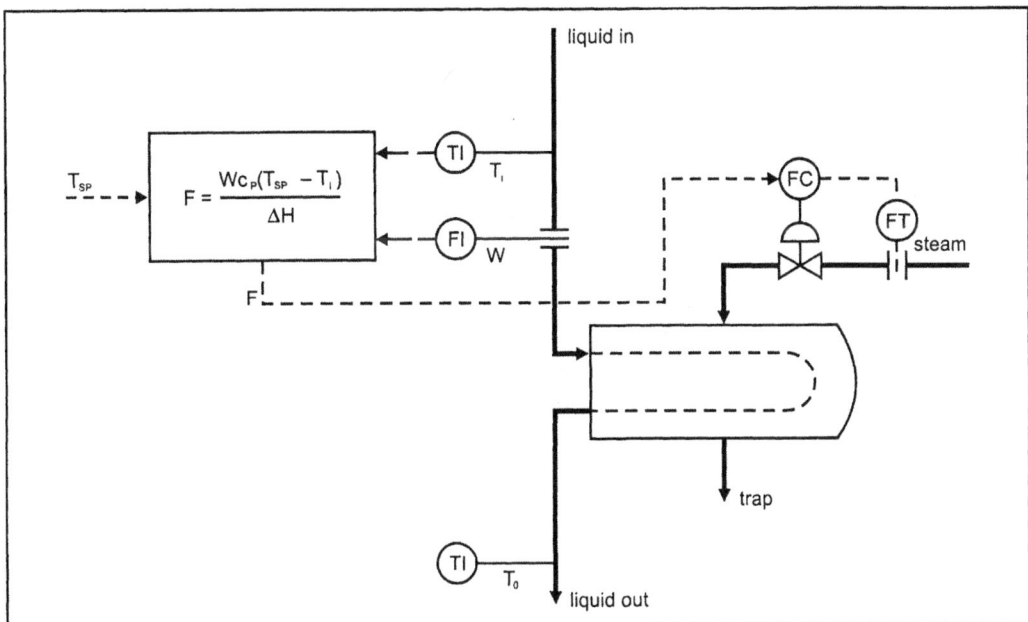

Figure 11-2. Feedforward Control of a Heat Exchanger

A steady-state energy balance around this exchanger gives the following equations:

$$WC_p(T_o - T_i) = F\Delta H \qquad (11\text{-}1)$$

where:

W = liquid flow, lb/hr
C_p = liquid heat capacity, Btu/lb – °F
T_i = liquid inlet temperature, °F
T_o = liquid outlet temperature, °F
F = steam flow, lb/hr
ΔH = heat released by steam, Btu/lb

Equation (11-1) can be solved for F:

$$F = \frac{WC_p(T_o - T_i)}{\Delta H} \qquad (11\text{-}2)$$

In this basic equation for F, the controlled variable T_o appears. Replace T_o with its set point or desired value T_{Sp} and the result is an equation that gives the steam flow needed to produce the desired outlet temperature:

$$F = \frac{WC_p(T_{sp} - T_i)}{\Delta H} \qquad (11\text{-}3)$$

This particular equation has W and the inlet temperature T_i measured and sent to the feedforward controller. Reasonable values for C_p and ΔH should be readily available and are entered by the operator. With this information and the desired value of the outlet temperature T_{Sp} it is possible to calculate the steam flow F. Feedforward control can then be implemented as shown in Fig. 11-2.

This particular example illustrates that the feedforward control equation for a specific loop is uniquely designed. Note the significant contrast to feedback control where standard control equations are normally constructed as combinations of three specific modes. To be able to design a particular feedforward algorithm for a particular process control application requires that the designer have significant levels of understanding. In addition, the use of feedforward control also means that the controller hardware used in feedforward control will be unique in each installation.

If you inspect Fig. 11-2 you will see that if there are errors involved in the computation or if there are other, unforeseen disturbances at work, the control hardware will have no way of sensing the resultant change in outlet temperature of the liquid. As a result, no effective corrective action will be taken.

11-3. Steady-State or Dynamic Feedforward Control?

The example in Fig. 11-2 was a specific application of steady-state feedforward control, that is, when the basic energy balance was made around the heat exchanger it was a steady-state energy balance. It did not give any quantitative recognition of the fact that there are process dynamics associated with the operation of the exchanger. But in practice, when the inlet conditions change the resultant changes in the controlled variable will not occur instantaneously. Yet in the specific feedforward control equation it was implicitly assumed that instantaneous correction was appropriate. When any condition changed in the feedforward control loop, there was an instantaneous (algebraic) calculation of a new steam flow.

To increase accuracy in feedforward control, it is desirable and often necessary to include the effects of process dynamics in the feedforward adjustment of the manipulated variable. This can be done by using one of two general strategies. In one case, it is possible to implement steady-state feedforward control and simply *push the output signal* through some type of dynamic compensation before that signal is used to adjust the manipulated variable itself. This type of dynamic compensation is illustrated in Fig. 11-3.

With dynamic compensation of the type illustrated in Fig. 11-3, it is possible to manually adjust or tune the dynamic compensation so that it introduces proper dynamic corrections to the feedforward control action that was taken. In typical hardware applications, quite often the dynamic compensation is a simple lag-lead network or a simple second-order lag adjustment. In such cases, the practitioner tunes the ratio of the lag-lead time constants or tunes the ratio of the time constants in the second-order lag network. Such tuning is done after the hardware is installed.

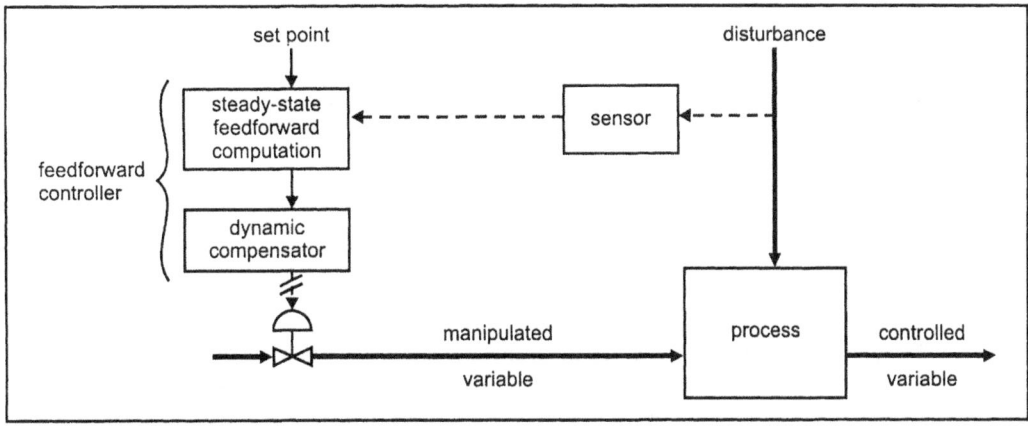

Figure 11-3. Dynamic Compensation

The second, and more theoretically sophisticated, way to provide dynamic compensation is through the general derivation of the feedforward control equation as a dynamic or unsteady-state relationship. This will be illustrated in the next section.

11-4. General Feedforward Control

It is possible to structure feedforward control in a more general way than we have done previously. Refer to Fig. 11-4, which shows a process to be controlled using a feedforward approach.

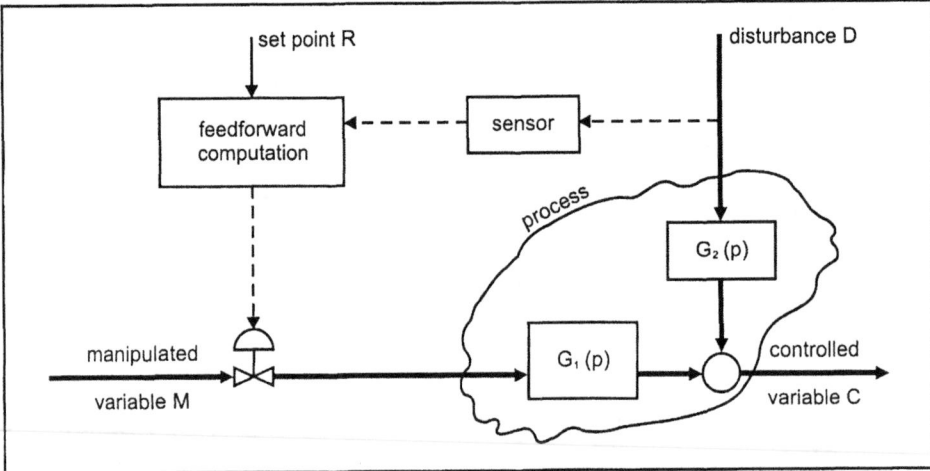

Figure 11-4. Generalized Feedforward Control Block Diagram[1]

If you look at Fig. 11-4, you can see that the controlled variable may be expressed in terms of the two input forces that are working on the process, that is, the disturbance D and the manipulated variable M. This relationship follows:

$$C = G_1(p)M + G_2(p)D \qquad (11\text{-}4)$$

This may be solved for the manipulated variable M:

$$M = \frac{C - G_2(p)D}{G_1(p)} \qquad (11\text{-}5)$$

---

1.   The symbol $G(p)$ implies the block diagram symbol for the dynamic relationship between the input to the block and the output from the block. The symbol p is the differential operator that implies d/dt, that is, it implies taking the derivative with respect to time. The implication is that the overall relationship may be a differential equation.

But in every case of pure feedforward control, you do not actually measure and feed back the controlled variable C. Instead, the reference or desired value R is taken and substituted for C in the overall control equation. This gives the following:

$$M = \frac{R - G_2(p)D}{G_1(p)} \tag{11-6}$$

This is the general feedforward control equation. D is measured, and the set point R is provided. We must provide the specific relationships $G_1(p)$ and $G_2(p)$ in advance. A continuous calculation is then made of the required value for M that will maintain $R = C$. The significant factor, however, is that the designer of this control system must have enough understanding of the process to be able to make a specific quantitative determination of the mathematical relationship between D and C, that is, $G_2(p)$, and the mathematical relationship between the manipulated variable M and the controlled variable C, that is, $G_1(p)$. You can clearly see the major increase in theoretical understanding required to implement feedforward control.

It should be understood that if two significant disturbances enter the process, each of them must have its own specific feedforward control computation, and, by implication, there will be an additional control equation similar to Equation (11-6). The difficulties involved in handling multiple disturbances make feedforward control impractical to implement on a very broad scale.

11-5. Combined Feedforward and Feedback Control

We have seen that pure feedforward control has some significant disadvantages, namely, the computations must be sophisticated and take into account many effects. Sometimes these computations involve sophisticated mathematical relationships. When errors are made in modeling or computation, there is no corrective action. In addition, if disturbances other than the specific ones being measured for the feedforward controller enter the process, the automatic control system does nothing and errors build up. As a result of these several problems, pure feedforward control is never encountered by itself. There is always some sort of feedback control added to supplement the feedforward arrangement. Several examples will illustrate how this is done.

Fig. 11-5 shows the simple heat exchanger of Fig. 11-2 as well as two ways to couple feedback control with the feedforward arrangement. In Fig. 11-5(a), a feedback controller is used to bias the output of the feedforward control output. In effect, a constant (or *fudge* factor) K_f can be added to our basic computation for steam flow. This gives the following:

$$F = WC_p(T_{Sp} - T_i)/\Delta H + K_f \tag{11-7}$$

In Fig. 11-5(a), the feedback controller will manipulate the value of K_f that is necessary to maintain $T_o = T_{Sp}$. Fig. 11-5(b) shows an alternate way to use a feedback control arrangement to supplement the feedforward control of the heat exchanger. In this particular case, the output of the feedback controller is used to adjust the set point of the feedforward controller.

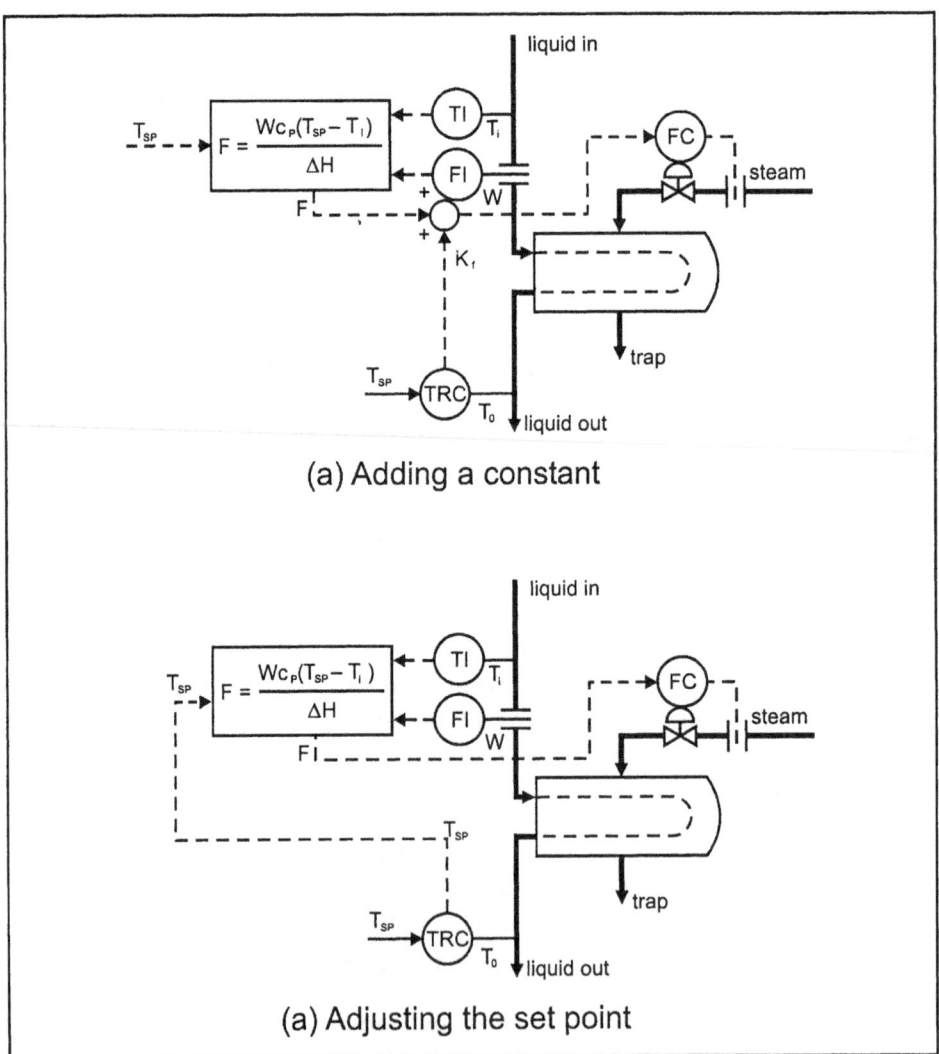

(a) Adding a constant

(a) Adjusting the set point

Figure 11-5. Two Ways to Use Feedback Control with Feedforward Control for a Heat Exchanger

The examples in Fig. 11-5 illustrate feedforward control combined with feedback control as well as a cascade arrangement to provide the control of the manipulated variable. This cascade aspect of the control scheme is common when feedforward control is used. It grows out of a general acceptance of the fact that if the variable is important enough to merit feedforward control, it is also inherently important enough to merit a cascaded arrangement to ensure that the manipulated variable is maintained at its calculated value.

One of the most frequently encountered examples of combined feedforward and feedback control is the control of water level in a boiler steam drum. This is illustrated in Fig. 11-6, which shows one of the classic three-element regulators.

Steam flow from the steam drum is measured, and this load on the system is sensed and compensated for by using feedforward control. Feedback control on the water level in the drum is handled in the conventional manner, and the manipulated flow into the steam drum (makeup water) is controlled through a cascade arrangement. This general arrangement is feedforward plus feedback plus cascade control--hence, the name "three-

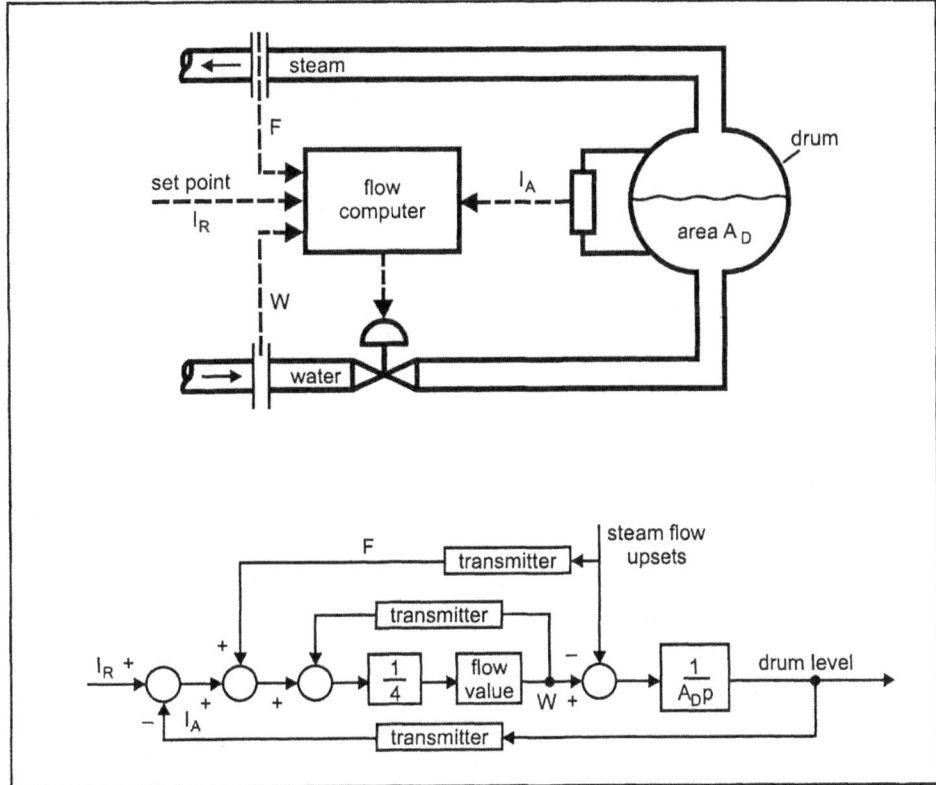

Figure 11-6. A Three-element Feedwater Regulator

element feedwater regulator." Fig. 11-6 shows a classic example of thistype of control, and there are literally thousands of these three-element regulators installed today.

In such a three-element arrangement, feedforward control in effect is used to provide compensation for significant variations in the major disturbance or the major load on the process. In arrangements like this (and Fig. 11-6 is a good example), many practitioners refer to the overall control arrangement as load compensation.

11-6. The Multivariable Control Problem

In previous units we have viewed the control system as being 1 × 1, that is, there was a single controlled variable and a single manipulated variable for the process.   But, in practice, this is seldom the case. Quite often, a process is found that involves more than one controlled variable and therefore more than one manipulated variable. This is illustrated in Fig. 11-7 for a 2 × 2 process (two controlled variables and two manipulated variables).

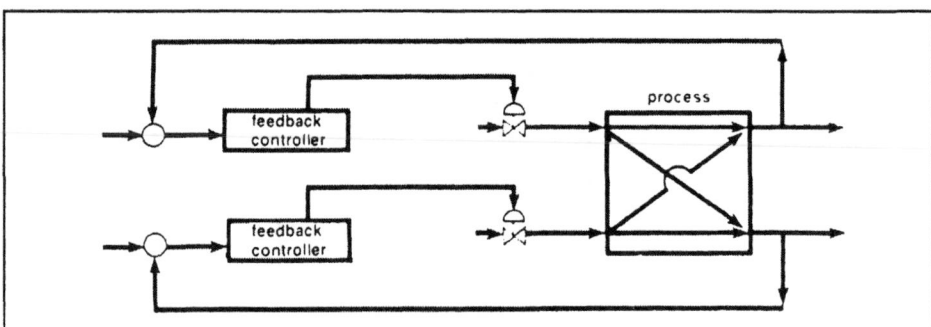

Figure 11-7. Conventional Multivariable Arrangement

When there are two control loops on one unit, quite often the two control units will *interact* with one another, that is., when there is a change in the manipulated variable in loop 1, not only will it produce a change in the controlled variable for loop1, it will also produce a change in the controlled variable for loop 2. Conversely, when there is a change in the manipulated variable for loop 2, not only will it produce a change in the controlled variable for loop 2 but in the controlled variable for loop 1 as well. This type of control interaction produces significant operating problems, which are illustrated in Fig. 11-8.

In Fig. 11-8, when there is a change in the manipulated variable for loop 1, there will be a change in the controlled variable for loop 1. However, in addition, the change in the manipulated variable for loop 1 will produce a change in the controlled variable for loop 2. This, of course, will be sensed, fed around loop 2, and will, in turn, produce a change in the manipulated variable for loop 2. This not only produces the change in the controlled variable for loop 2 but also produces a further change in the controlled variable for loop 1. In effect, these phenomena start to chase one another in a figure-eight type of path. To compensate for this, the operator must desensitize the loops, and it becomes very difficult to obtain quality feedback control.

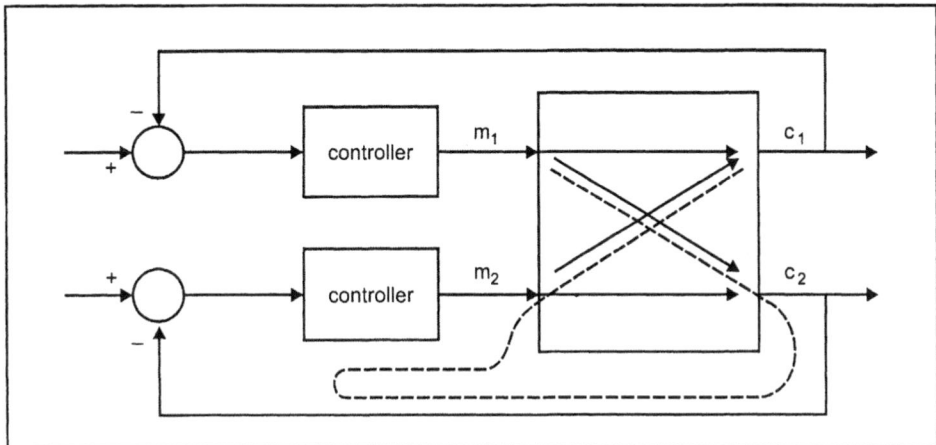

Figure 11-8. The Effects of Interaction in a 2 × 2 System

11-7. Implementing Multivariable Control

When feedback loops are interacting with one another, a control system is needed that will *decouple* the loops; that is, the interaction between the loops needs to be broken. The concept of breaking this interaction is referred to as *multivariable control*. The basic approach is to use conventional feedback control supplemented by a *decoupler*. This is illustrated in Fig. 11-9.

In effect, a decoupler like that shown in Fig. 11-9 is essentially two feedforward control elements. The overall operation is such that when a change in the controller output for loop 1 occurs, it not only produces a change in the manipulated variable for loop 1. It also produces a change in the manipulated variable for loop 2. Thus, when the interaction from loop 1 to loop 2 takes place, the change in the manipulated variable in loop 2 will compensate exactly. As a result, there will be no change in the controlled variable for loop 2. Conversely, in loop 2, when there is a change in the controller output it produces a change in the manipulated variable for loop 2, but it also produces a change in the manipulated variable for loop 1. Thus, when the interaction from loop 2 to loop 1 takes

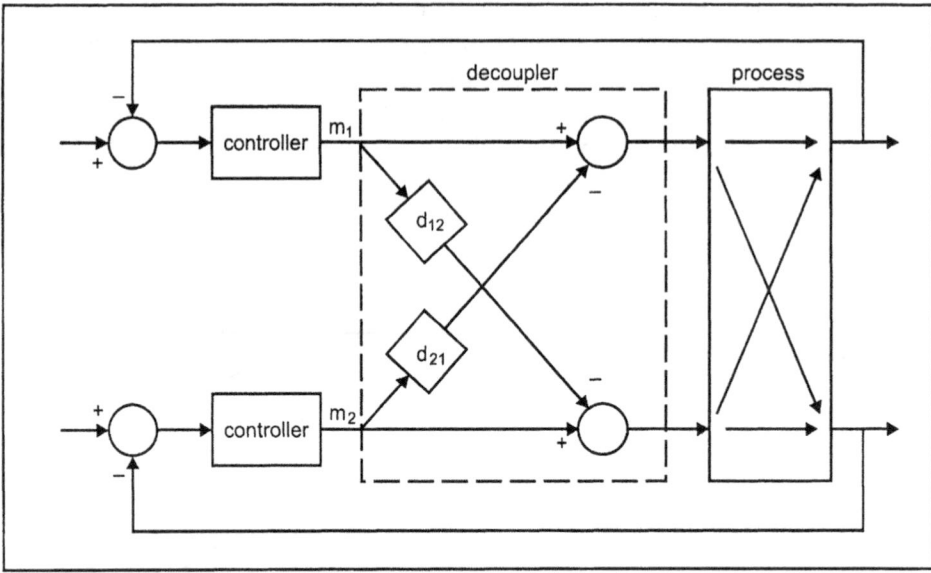

Figure 11-9. A Decoupler for a 2 × 2 Process

place, it will be compensated for by the change in manipulated variable in loop 1, and there will be no change in the controlled variable in loop 1.

The concept entailed by the decoupler shown in Fig. 11-9 is very appealing. The difficulty is in the design of the decoupler itself. The goal is to design the decoupler so that all interactions will be compensated for exactly. This is a laudable goal but often difficult to achieve in practice. For a 2 × 2 process, it is equivalent to designing two simultaneous feedforward controllers. The problem escalates rapidly, however, and if the system were a 5 × 5 process the implementation of such a multivariable control would require the implementation of 5^2 - 5, or 20 simultaneous feedforward controllers. The increasing difficulty is obvious.

Quite often a decoupler such as that shown in Fig. 11-9 is implemented as a steady-state decoupler, that is, one consisting of algebraic terms only. It is then possible to add dynamic compensation to the more important elements of the decoupler later and gain further improvement.

You should begin to suspect by now that multivariable control represents a most powerful and most difficult type of automatic control to implement. It requires significant design expertise and capability, and, as a result, it also requires custom hardware capabilities. This conventionally implies the necessity of having digital computation capabilities in the hardware. The cost of designing for multivariable control can be quite significant, but the improvement in control also can be significant. There are increasing numbers of commercial applications of multivariable control that have excellent economic justification.

EXERCISES

11-1. *Refer to Fig. 8-5. Design a steady-state feedforward control system for this unit.*

11-2. *To Exercise 11-1, add a lag-lead dynamic compensation of the following form:*

$$\frac{1 + \tau_1 p}{1 + \tau_2 p}$$

where the ratio of τ_1/τ_2 is adjustable (tunable).

11-3. *Add cascade control of the manipulated steam flow to the drawing in Exercise 11-2.*

11-4. *Refer to the results of Exercise 10-4. Install a flowmeter on the feed input to the column, maintain the reflux flow rate as a ratio of column feed flow, and send a signal from the feed orifice through the ratio controller and then through a dynamic compensator. Now use the composition controller to adjust the ratio.*

11-5. *Given the following 2 × 2 control system,*

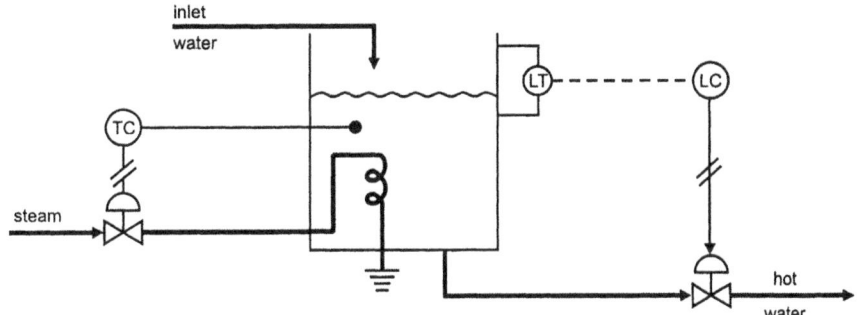

draw the functional block diagrams for these two loops.

11-6. *Sketch the functional outline of a decoupler for the system of Exercise 11-5.*

Unit 12:
Special-Purpose
Concepts

UNIT 12

Special-Purpose Concepts

There are a number of important variations, extensions, and modifications of feedback and feedforward control. To illustrate the diversity of possible concepts, in this unit we will present several such implementations that while important do not merit separate units.

Learning Objectives — When you have completed this unit, you should:

A.   understand the use of computing blocks in control system design

B.   know how to apply ratio control

C.   understand the concept of override control

D.   understand the concept of selective control

12-1. Computing Components

As we employ more advanced control concepts, we need to make more computations. Whether the computational power comes from computing relays, microprocessors, or other variations of hardware and/or software is not of primary importance to us in this book since we are focusing on the control concepts themselves. Obviously, these functions are now done primarily by microprocessors. However, as we think about various control schemes it is helpful to think about the manipulations that are involved and to consider the functional character of various computational components. A typical list of such computational components includes the following:

1.   *Addition/subtraction.* The output signal is either the algebraic sum or difference of the input signals.

2.   *Multiplication/division.* The output signal is produced by either multiplying or dividing the input signals.

3.   *Square root.* The output signal is the square root of the input signal.

4.   *High/low selector.* The output signal is the highest/lowest of two or more input signals.

5.   *High/low limiter.* The output signal is the input signal limited to some present high/low limit value.

6. *Function generator.* The output signal is a function of the input signal.

7. *Integrator.* The output signal is the time integral of the input signal. An integrator is often called a totalizer.

8. *Linear lag.* The output signal is the solution of a first-order differential equation in which the input signal is the forcing function. This is expressed mathematically as follows:

$$output = \left(\frac{K}{1 + \tau p}\right) input$$

where K is gain, τ is the time constant, and p is the Heaviside operator d/dt.

9. *Lead-lag.* The output signal is the solution to the following differential equation:

$$output = K\left(\frac{1 + \tau_1 p}{1 + \tau_2 p}\right) input$$

where K is gain, τ_1 is the lead time constant, and τ_2 the lag time constant, and p is d/dt. This is often used in dynamic compensation.

The use of microprocessors extends this list virtually without limit. Complex calculations can be done easily and without major hardware or software limits. The most common limit is the insight and skill of the practitioner. We will discuss these computation components more and more throughout the balance of this text.

12-2. Ratio Control

A commonly encountered type of multiple-loop feedback control system is *ratio control* (sometimes called *fraction control*). When one looks at just the hardware, ratio control quite often is confused with cascade control because in ratio control one loop adjusts another. However, the operation of ratio control is quite different from cascade control.

Ratio control is the simplest form of feedforward control. You measure the disturbance (the wild flow) and determine by way of a model (the ratio) the value of the manipulated variable. Ratio control is often associated with process operations in which it is necessary to mix two or more streams together continuously to maintain a steady composition in the resulting mixture. A practical way to do this is to use a conventional flow controller on one stream and control the other stream with a ratio controller that maintains that stream flow at some preset ratio or fraction to the primary stream flow.

A ratio control system for regulating the composition of a feed stream to a reactor is shown in Fig. 12-1. (According to ANSI/ISA-S5.1, *Instrumentation Symbols and Identification*, the FFY symbol denotes flow rate (first F), Ratio or Fraction (second F), and functions associated and defined outside the circle (the Y). See Appendix A.) The block diagram arrangement for this system is shown in Fig. 12-2. In this diagram, the subscript 1 refers to the air stream in Fig. 12-1, while the subscript 2 refers to the hydrocarbon stream. R is the ratio station that is adjustable. The ratio element is actually just a multiplier with an externally adjusted gain.

The design of a ratio control system poses no special problem because each of the loops is designed individually. It is also possible to implement a ratio control system if the primary instrument is simply a transmitter. In such a situation, the set point of the controller is set in direct relation to the magnitude of the primary controlled variable. An example of this is shown in Fig. 12-3, and the associated block diagram for this type of ratio system is shown in Fig. 12-4. Basically, the general principles are very similar to those of other ratio control systems, except one of the streams is uncontrolled, and the other is simply maintained in ratio to it.

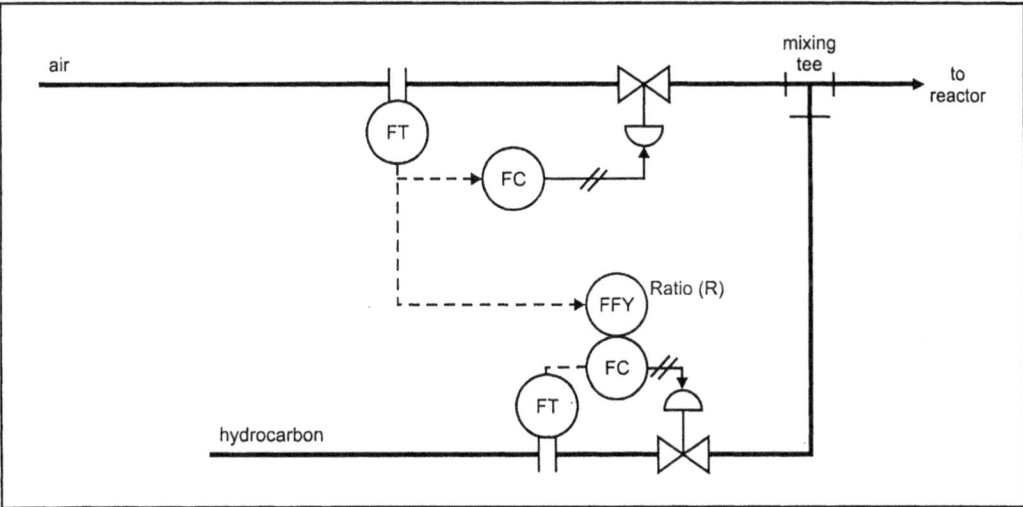

Figure 12-1. Conventional Ratio Control System

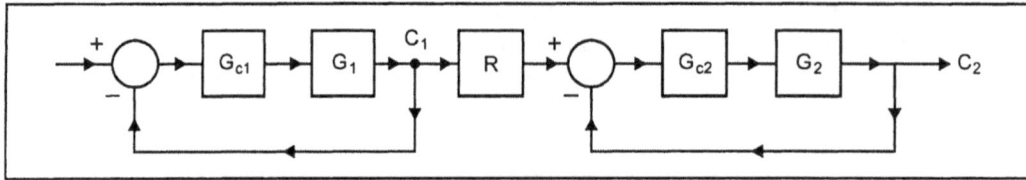

Figure 12-2. Conventional Ratio Control System's Block Diagram

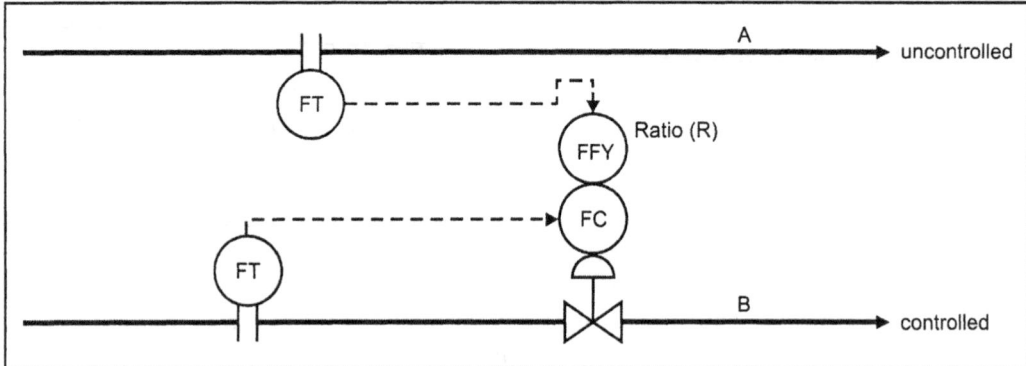

Figure 12-3. Ratio Control with an Uncontrolled Stream

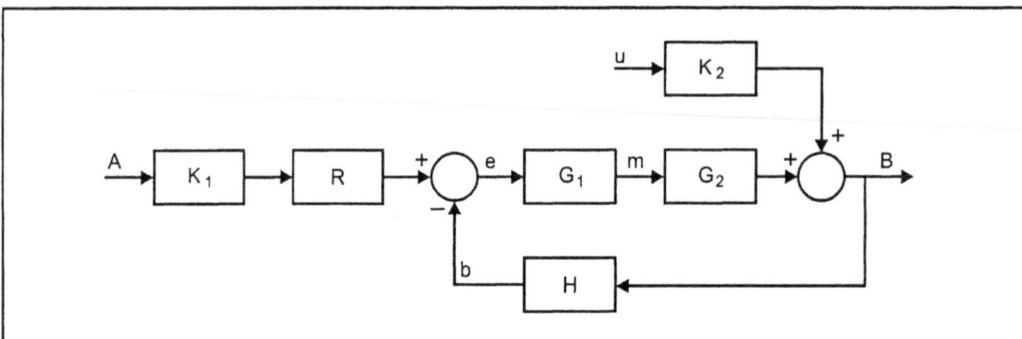

Figure 12-4. Block Diagram for Ratio Control with an Uncontrolled Stream

12-3. Applying Ratio Control

You can get a much better grasp of ratio control by considering some specific examples. Assume we want a "blender," that is, to blend two streams, A and B, in a proportion or ratio R:

$$R = B/A \qquad (12\text{-}1)$$

Two simple schemes for doing this are shown in Fig. 12-5. In Fig. 12-5(a), the wild flow A is measured, and then it is multiplied by R to get the required value of B. Thus, as the flow A varies, the set point to the flow

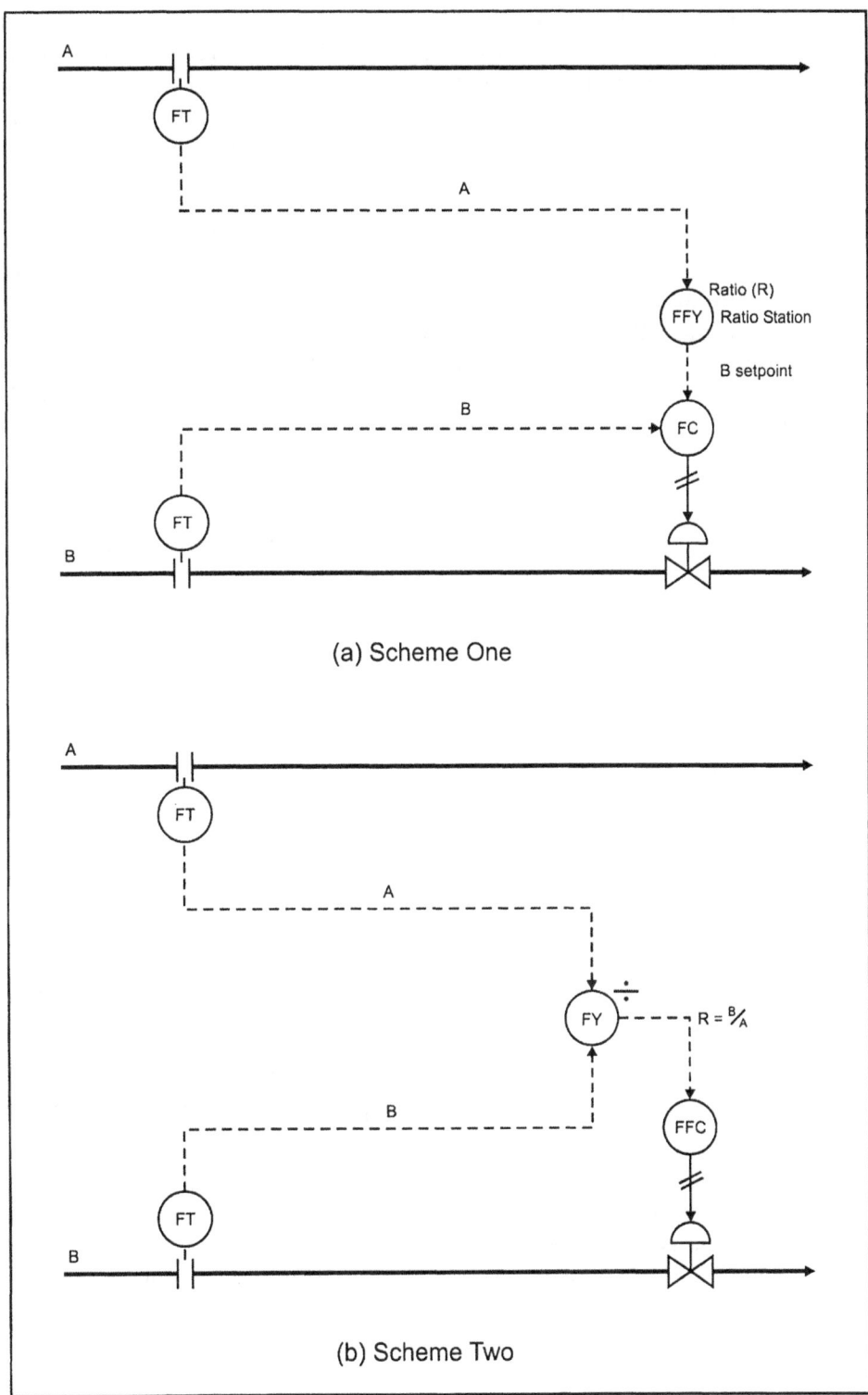

Figure 12-5. Two Ratio Control Systems

controller of stream B will vary to maintain R. If a new value of *R* is desired, it must be set into the ratio station. Differential-pressure sensors are shown measuring flow; their output indicates the square of flow and, therefore, square root extractors are shown to obtain flow. By using flow, and not its square, the loops will behave more linearly and thus will be more stable and easier to tune.

In Fig. 12-5(b), both streams are measured, then divided to obtain the actual ratio. The actual value of R is then sent to a ratio controller where it is compared with a set point value of R, and the difference is used to manipulate B. Both of the schemes of Fig. 12-5 are used in industry, but the arrangement shown in Fig. 12-5(a) is preferred because it is more linear than Fig. 12-5(b). We can show this increased linearity. First, consider Scheme 1 where the ratio station calculates the following:

$$B = RA$$

The gain of the ratio station is the amount the output changes for a change in input:

$$\frac{\partial B}{\partial A} = R \qquad \text{(a constant)}$$

Now consider Scheme 2:

$$R = B/A$$

$$\frac{\partial R}{\partial A} = -\frac{B}{A^2}$$

Thus, in Scheme 2, when flow A changes, the gain changes, and a nonlinearity is present that makes the system less stable and more difficult to control.

As a specific example of ratio control, consider the control of the air-fuel ratio to a steam boiler, as shown in Fig. 12-6. This is actually called *parallel positioning control* because we are really maintaining a ratio to the final control elements. A better control scheme is to establish *full metering control*, as shown in Fig. 12-7, where the fuel flow is set by the pressure controller, and the ratio station sets the air flow rate as a ratio of fuel flow rate. The flow loops will correct for any flow disturbances.

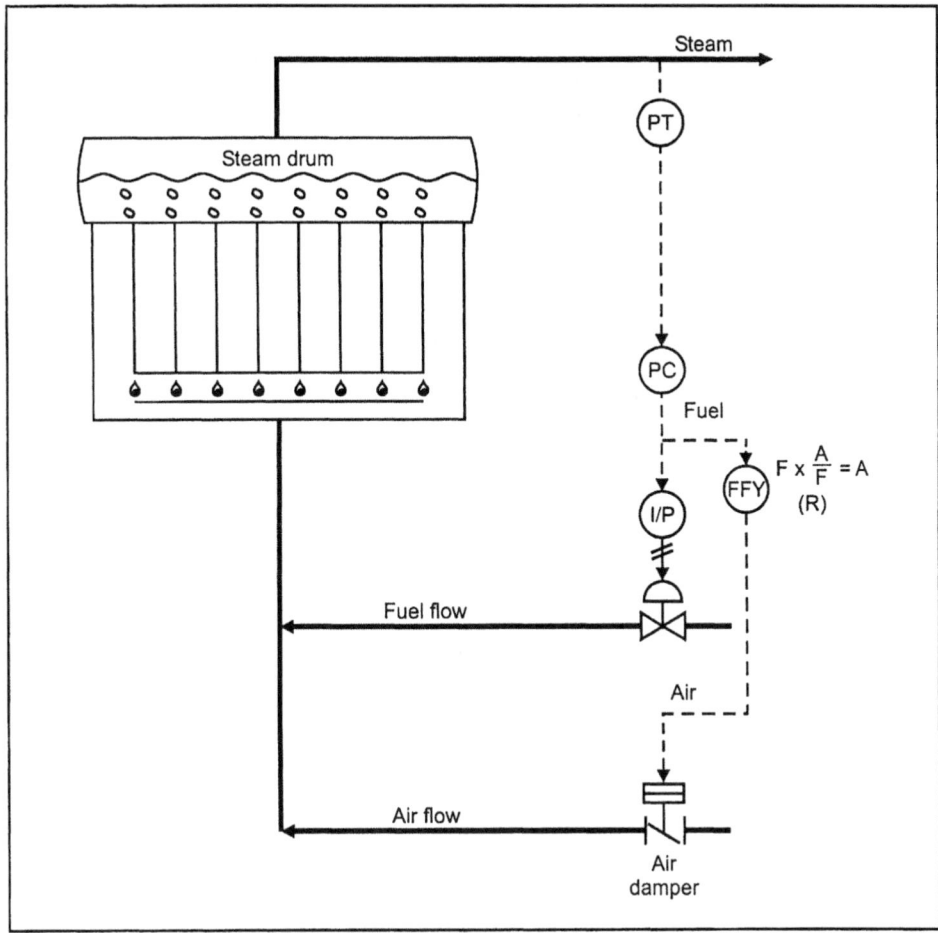

Figure 12-6. Air/Fuel Ratio Control in a Steam Boiler

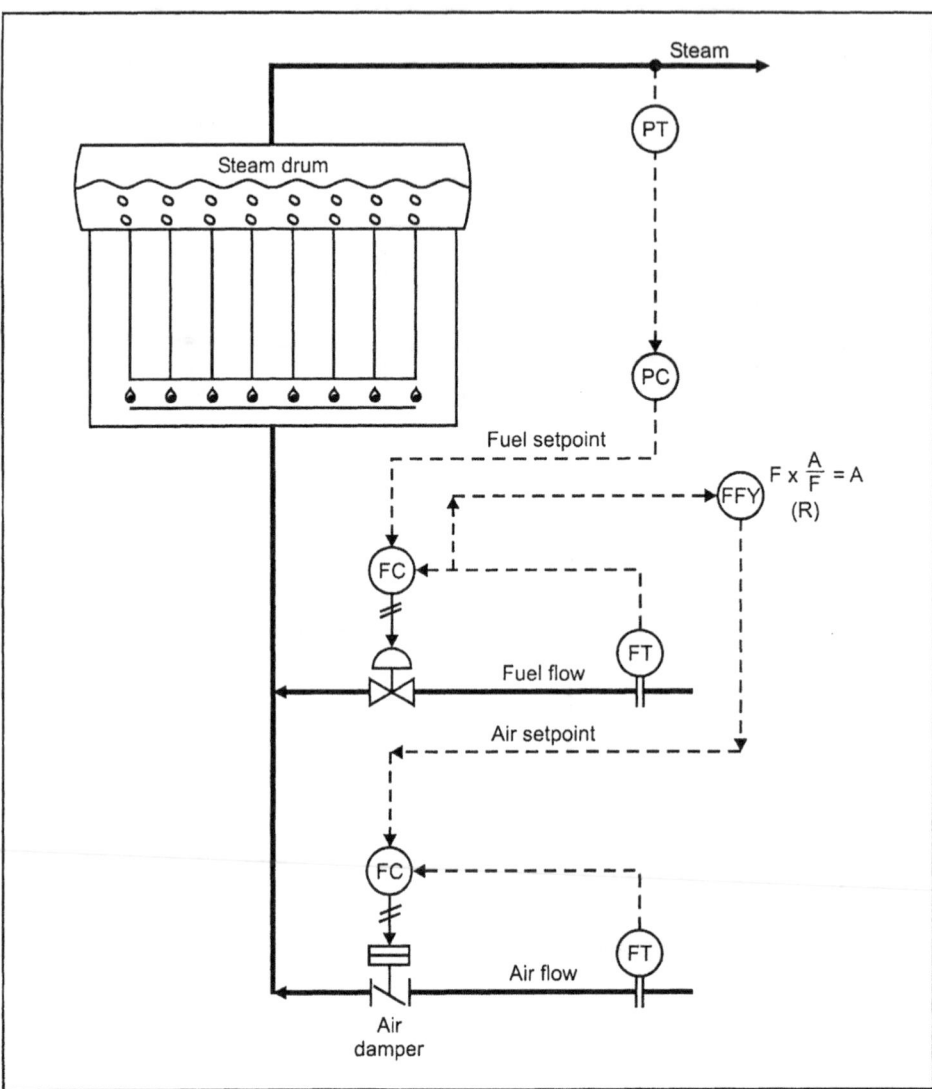

Figure 12-7. Full Metering Control

This ratio control concept can be carried further by using a stack gas analyzer to trim the ratio of air to fuel. This is shown in Fig. 12-8.

Several comments are helpful to extend this particular application. High and low limiters are usually used with control schemes like those as Fig. 12-8 to ensure safe operations. Also, a concept called *cross-limiting control* is normally used to ensure that during transients the combustible mixture is always enriched in air. Although the sketches presented so far have shown calculation and control functions separately, in digital installations these are often combined into a single microprocessor.

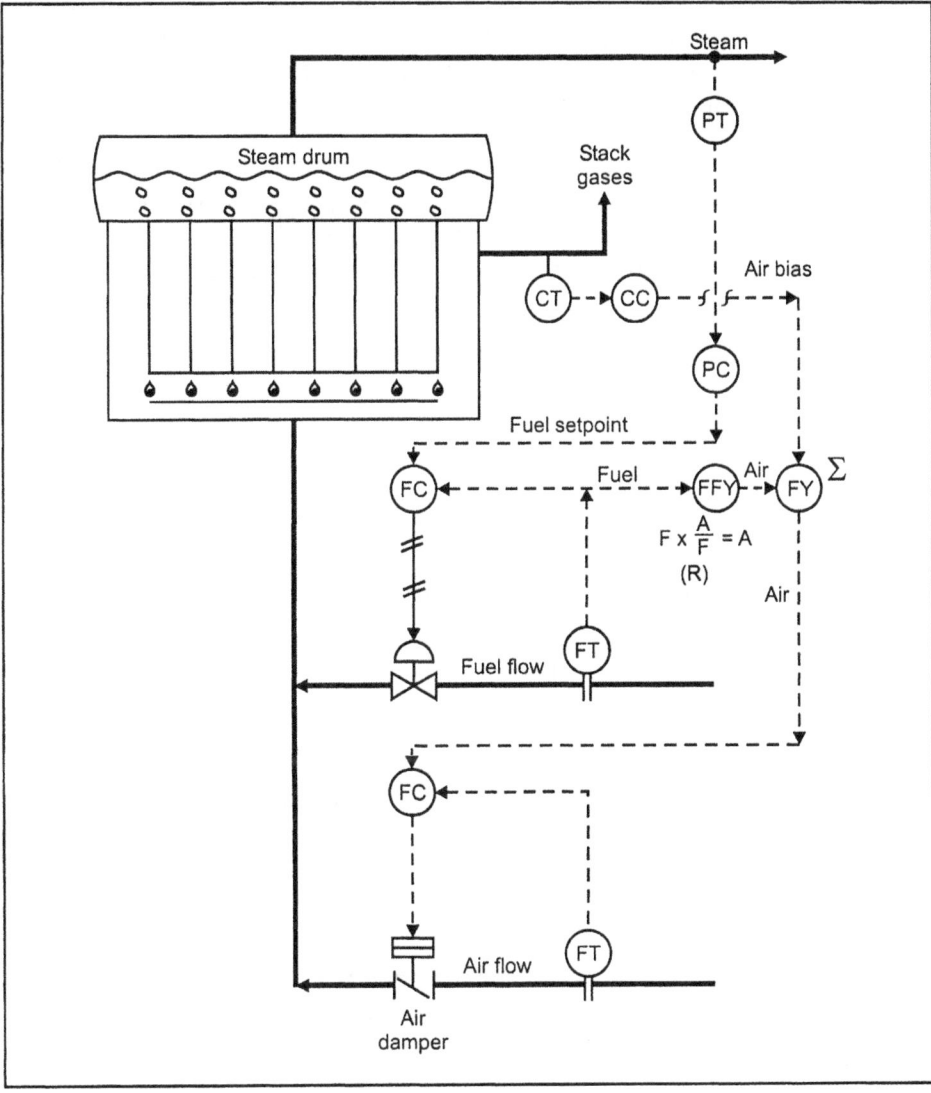

Figure 12-8. Cascade Control of the Ratio of Air Flow to Fuel Flow

12-4. Override Control

The *override control* concept is a technique by which process variables are kept within certain limits, usually for protective purposes. (*Interlock control* is an alternate protective scheme, but it is usually oriented toward equipment malfunction, and it acts to shut the process down. Override control is not so drastic.) Override control maintains the process in operation but within and under safer conditions.

To illustrate override control, consider the simple process shown in Fig. 12-9. A hot, saturated liquid enters a process surge tank, and then it is then pumped into the process. Normally, the tank operates at the level shown, but if the level gets too low the liquid will not have enough net positive suction head (NPSH), and the pump will start to cavitate. Override control can provide protection; a scheme for this is shown in Fig. 12-10.

Fig. 12-10 shows that in this override control application, the tank level is now controlled. The variable-speed pump will, of course, pump more liquid as the energy input to it increases. It follows that the flow controller must, therefore, be a reverse-acting controller (output increases as input

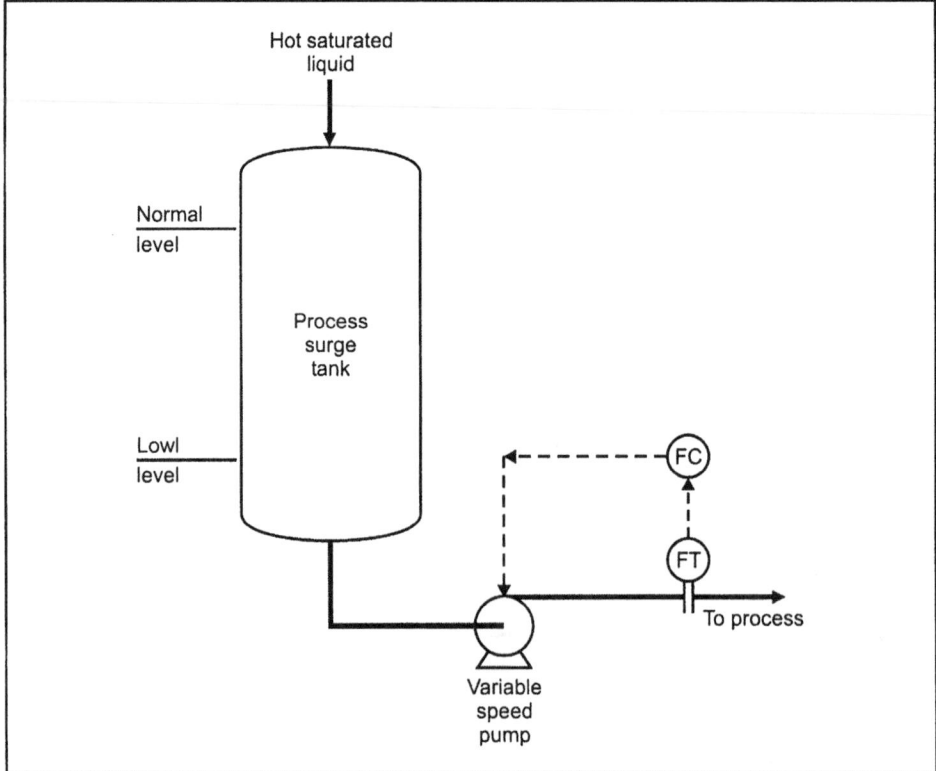

Figure 12-9. A System Needing Override Control

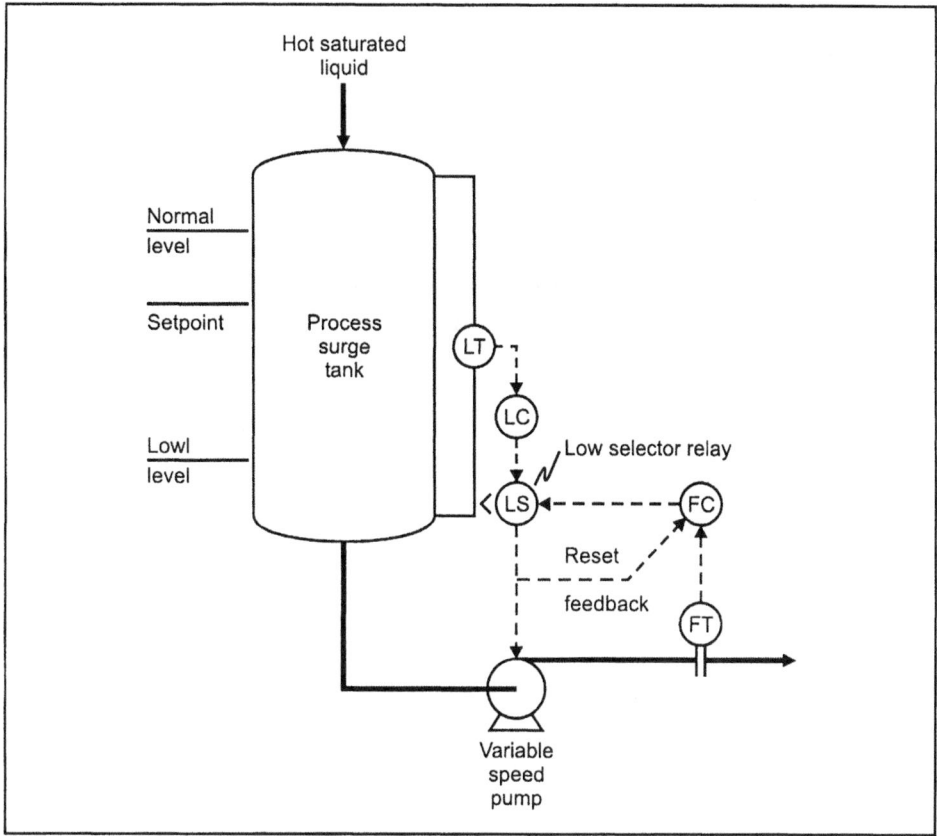

Figure 12-10. Override Control System

decreases), and the level controller must be a direct-acting (output increases as input increases) controller. The output of each controller is connected to a low-level selector relay, and its output goes to the pump.

Under normal operating conditions, the actual level is above the set point of the level controller, and the level controller will attempt to speed up the pump. Normally, the output of the flow controller will be less, and the low-level selector relay will select the flow controller output to manipulate pump speed. If the flow of hot, saturated liquid slows down and the level drops, the level controller will try to slow down the pump by reducing its output. When the output of the level controller drops below the output of the flow controller, the low-level selector relay will select the output of the level controller to control the pump. Now the level controller "overrides" the flow controller. Override control is also often called *constraint control*, for obvious reasons.

In override control, it is essential that any controller that has integral (reset) action must also have reset windup protection. This is shown in Fig. 12-10 for the flow controller. The reset feedback structure of Fig. 12-10 has the advantage of providing for the elimination of reset windup in override and cascade control systems in a very clean and straightforward manner. Without such protection, the controller action is delayed, pending "unwinding," and the override protection will probably be too late. In the example shown in Fig. 12-10, the flow controller will not wind up because of the reset feedback signal.

As another example of override control, consider distillation column differential-pressure control as shown in Fig. 12-11. Temperature control is normally used to provide manipulation of steam flow to the reboiler. But if the column experiences downcomer flooding, it will greatly increase column differential pressure, and steam flow rate must be reduced. The scheme shown in Fig. 12-11 allows the temperature controller to set reboiler steam flow rate as long as the column differential pressure is below its set point. This set point is the maximum desired differential pressure. In the example shown, both the temperature controller and the differential-pressure controller are assumed to have the reset mode, both can be overridden, and thus both need reset windup protection. In this example, neither the temperature controller nor the ΔPC controller will wind up because of the reset feedback signal. As long as one of the controllers is in control, the other controller cannot wind up.

There are many other examples of override control in use throughout the process industries. Its value is obvious.

12-5. Selective Control

Each controlled variable in a process must be paired with a specific manipulated variable at any given instant of time.   However, since the number of controlled variables and the number of manipulated variables may not be equal, a logical means of sharing variables among loops needs to be devised. It may be necessary to switch a controller from one controlled variable to another, or it may be necessary to switch a controller from one manipulated variable to another. This can be done with "selective" devices that have two or more inputs and produce a single output. Depending on need, this output may be highest, lowest, or the median of the inputs. The use of such selective devices is a way to achieve constraint control on flows and other operating conditions.

First, consider the case of constraining or choosing among manipulated variables when there is a single controlled variable. Refer to Fig. 12-12, which illustrates the fuel supply system for a process heater. Assume that the primary fuel A is burned up to the high limit of its availability at any moment, and then it is supplemented with fuel B. Fuel A has a high limit that is an input to a low selector on the set point of A. In Fig. 12-12, this is shown as a manual input of the high limit, but it could be provided

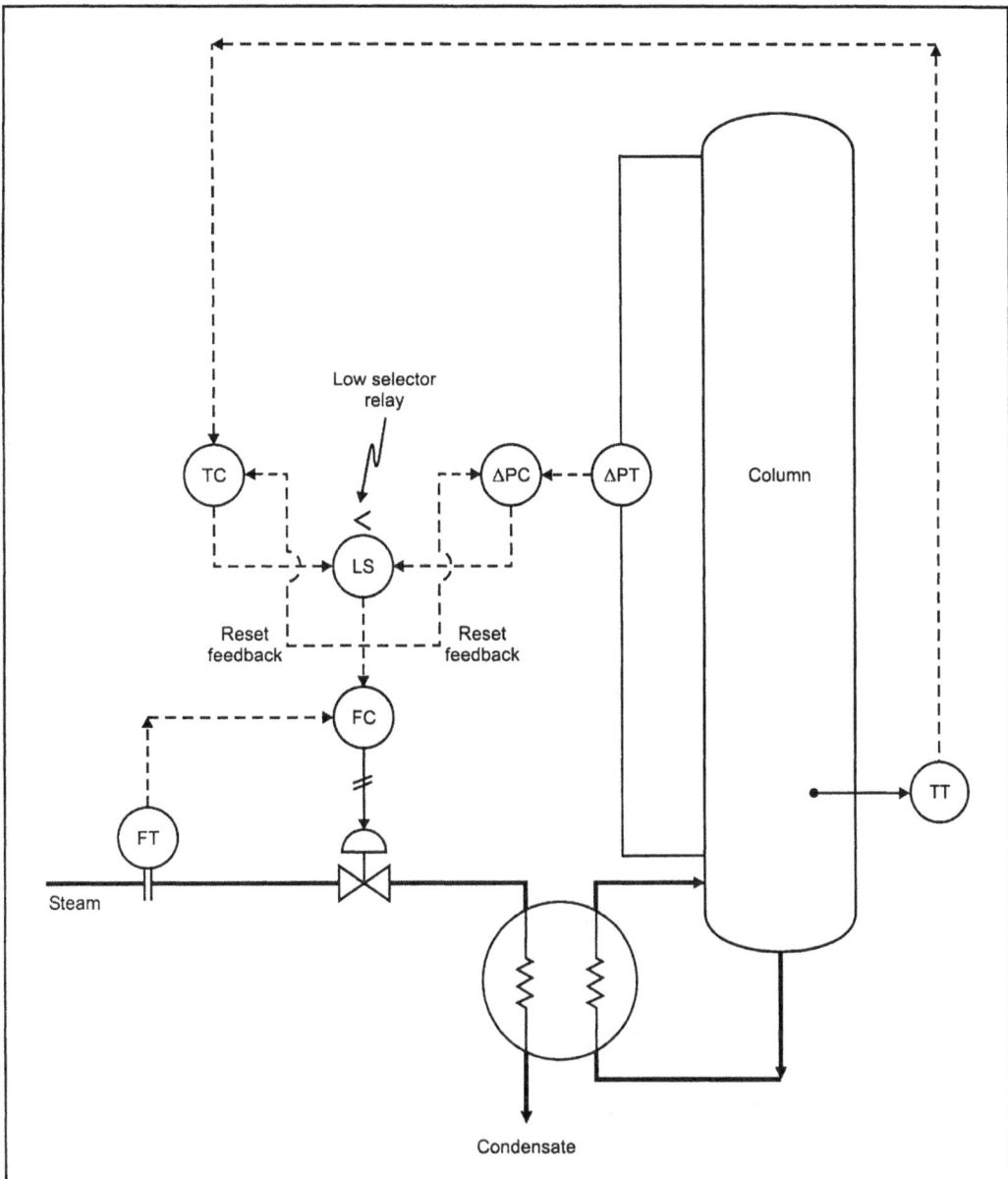

Figure 12-11. Column Differential Pressure Override Control

automatically. In Fig. 12-12, any difference between the output of the temperature controller and the flow of fuel A is converted to a set point signal for fuel B. In reality, the temperature controller output represents the total heat input to the system from the two combined fuels. This is converted into a set point for fuel A by multiplying by $(1 + B/A)$ where B/A is the ratio of full-scale heat input of fuel B to fuel A. This equates the heat-flow values of controller output and fuel A set point.

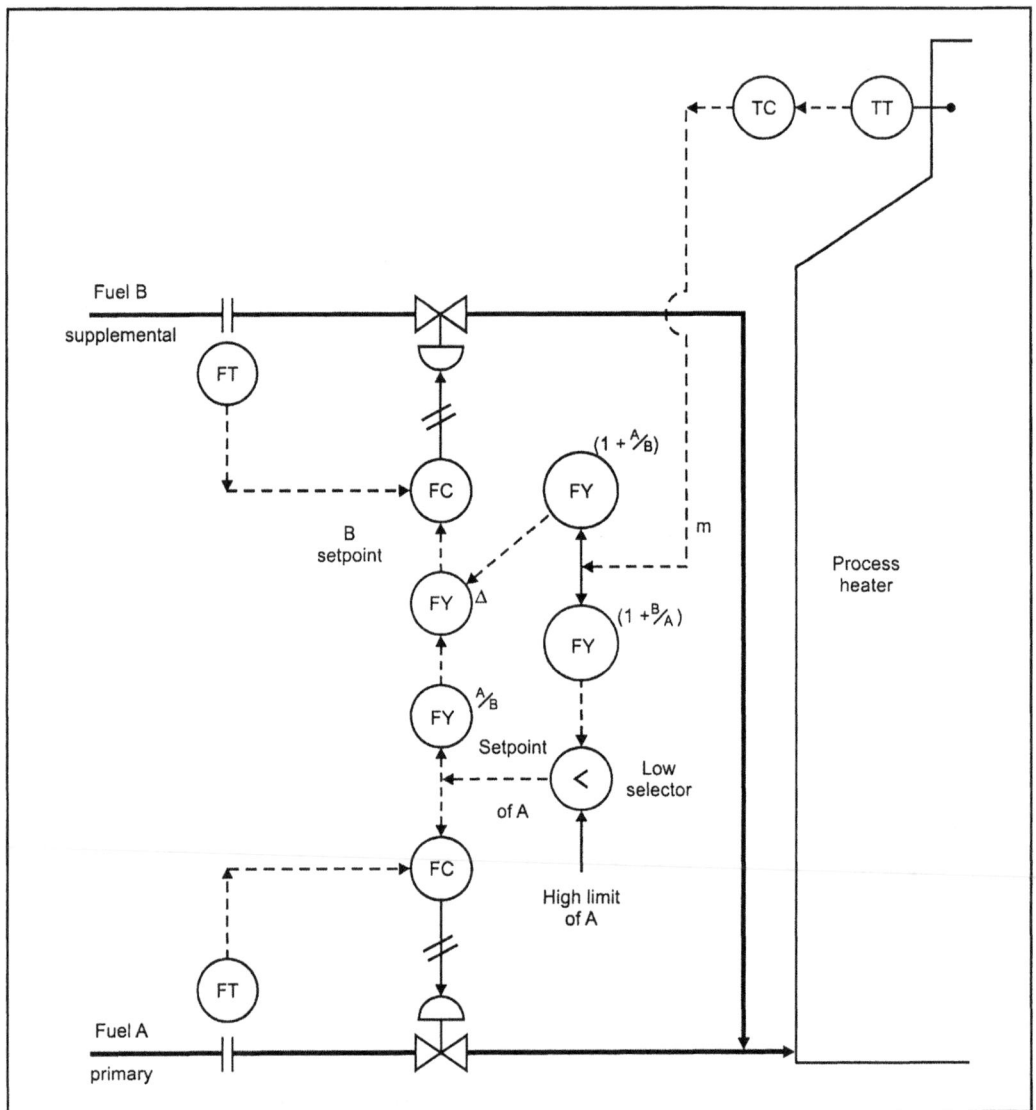

Figure 12-12. Example of Selective Control with Two Manipulated Variables and One Controlled Variable

When fuel A is at or below its high limit, the two inputs to the subtracting relay will cancel one another:

$$m (1 + A/B) - m (1 + B/A) (A/B) = 0 \qquad (12\text{-}2)$$

and the B set point will be zero. When the high limit or fuel A is exceeded, F_A becomes constant and F_B starts to increase.

In selective control, we use similar types of selecting devices and logic when limits must be placed on one or more controlled variables with a single manipulated variable. In such cases, the controller output is the point at which selection is made. Refer to Fig. 12-13 for a scheme in which pressure is controlled in a pipeline by a throttle valve on a pump discharge whenever the pump motor current is below its rated limit. If the pump motor current ever increases to its limit, the current controller IC takes over manipulation of the control valve from the pressure controller PC and closes the valve further. As a result, pipeline pressure is controlled below the set point on PC. (This can also be viewed as override control.)

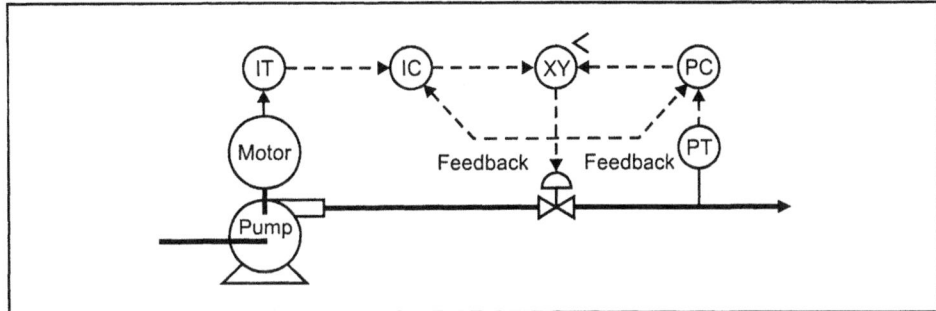

Figure 12-13. Selective Control of Pressure in Pipeline

In both of the examples discussed in this section, constraint control used low-signal selectors, and they are, in fact, more common than high-signal selectors. This is due to the fact that most final operators (usually control valves) fail safely on loss of signal, and, at the same time, most constraint control provides protection from the high values of variables.

12-6. Duplex or Split-range Control

The controller in duplex or split-range control has one input and two outputs. A good example of the application of duplex control is shown in Fig. 12-14. Assume that the temperature controller TC is a proportional-only controller; its output is fed to both control valves. If the valve signal varies from 3 to 15 psi, as is typical, the valve action would be set so that the steam valve traveled from full open to closed as input to the positioner went from 3 to 9 psi. Moreover, the water valve would go from closed to full open as the signal went from 9 to 15 psi. The system is designed so that the controller output produces a 9 psi signal when measurement and set point agree, and, at this point, both valves are closed. If measured temperature rises above or falls below the set point, either water or steam will be circulated in proportion to the difference between the set point and the measured value of temperature.

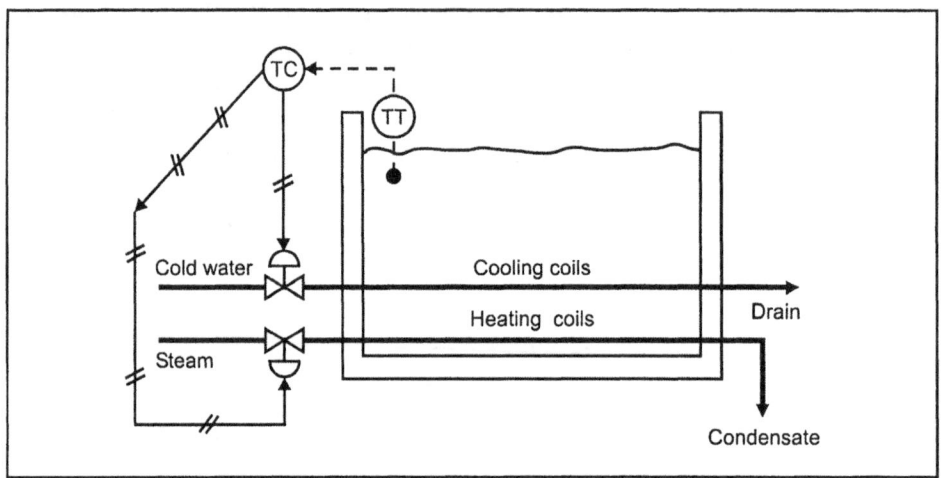

Figure 12-14. Duplex Control of Bath Temperature

Usually, for safety reasons, the valves in Fig. 12-14 should work in a way opposite to the split just described. The cooling valve would operate 3 to 9 psig, and the steam valve would operate from 9 to 15 psig. If a failure occurs and the controller output goes to zero, the system would then go to full cooling.

Duplex control is conceptually similar to selective control, as was shown in Fig. 12-12, which also used a single-controlled variable to constrain or choose between two manipulated variables.

12-7. Auto-Selector or Cutback Control

Auto-selector control is the opposite of duplex control in that it selects between two or more measurement inputs so as to provide a single output to control a single valve. It also is a form of selective control, similar to that shown in Fig. 12-13, in which there can be more than one controlled variable associated with a single manipulated variable. Some examples of auto-selector control are presented in the exercises at the end of this unit.

EXERCISES

12-1. *Diesel oil is injected into the suction of a fuel oil pump (the impeller does the mixing to control viscosity in fuel oil):*

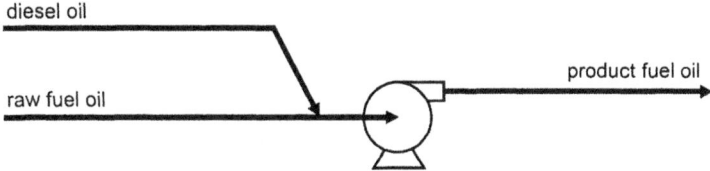

Design a ratio control system for this pump.

12-2. *Install a viscosity meter as a sensor on the product fuel oil line of Exercise 12-1 and use it to adjust the ratio between the two streams.*

12-3. *Consider the following control scheme for two flows into a mix tank:*

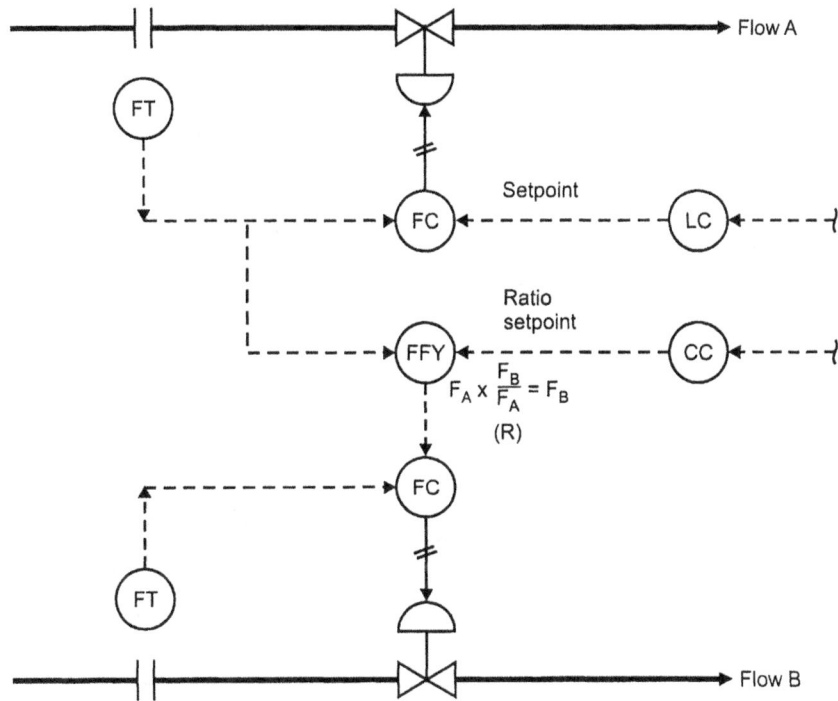

Why use the liquid-level controller to set both flows?

12-4. *Refer to Exercise 12-3. Which stream should be larger, flow A or flow B?*

12-5. *Consider the following ratio system:*

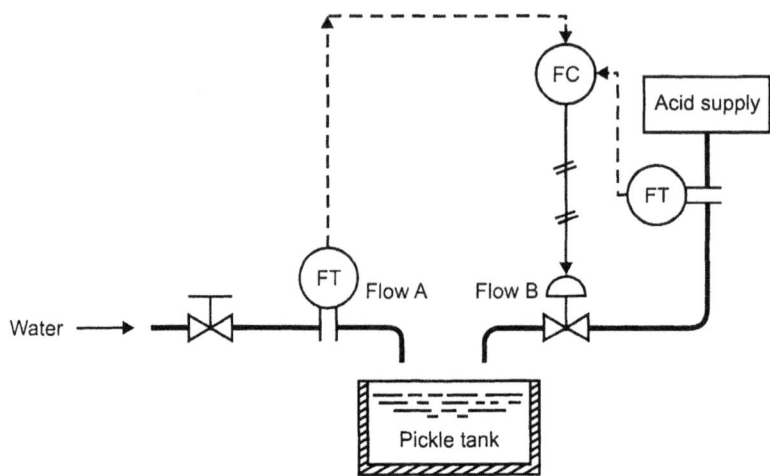

Show how this could be modified so that the ratio of flow A to flow B could be adjusted through a ratio station?

12-6. *Consider the following catalytic exothermic plug flow reactor, along with the typical temperature profile along the reactor:*

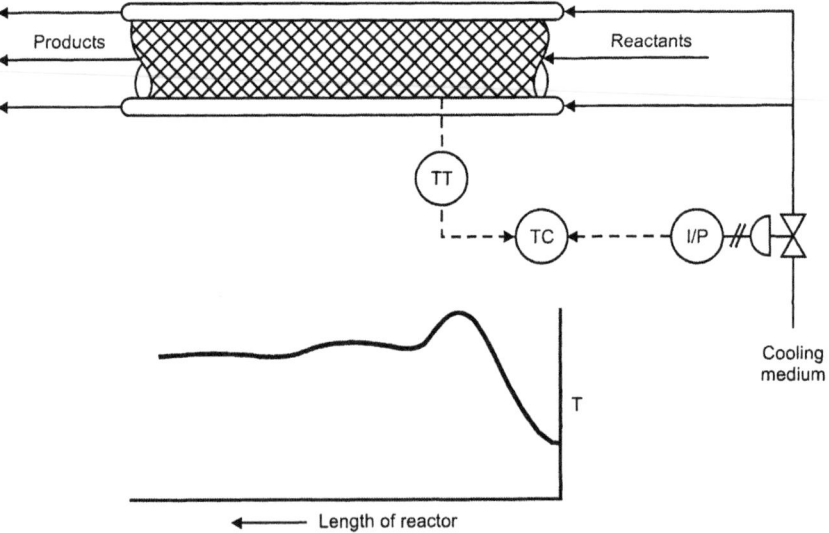

As conditions change and as the catalyst ages, the hot spot moves. Design a selective control scheme so that the measured variable moves as the hot spot moves. Use multiple temperature measurements with a high selector.

12-7. *To operate a pumping station on a pipeline efficiently, it is best to operate the control valve wide open on all stations but one. This station becomes the throttling unit. It is necessary to cut back on the control valve position if*

   a.   *suction pressure drops too low*

   b.   *motor load rises too high*

   c.   *discharge pressure rises too high.*

Sketch an auto-select/cutback control system to handle this.

Unit 13:
Dead Time Control

UNIT 13

Dead Time Control

Dead time has been called "the most difficult element to control." Not only is it difficult, but the control of dead time requires relatively sophisticated mathematics and digital hardware. Because of the difficulty and uniqueness of dead time control, it merits a special unit.

Learning Objectives — When you have completed this unit, you should:

A. understand the nature of a process dead time

B. understand the need to review control system design to eliminate unnecessary or inappropriate dead times

C. appreciate the use of predictor algorithms and techniques

13-1. Dead Time

Before discussing the control of dead time, it is helpful first to get a feel for its physical implications. To do this, refer to Fig. 13-1, which shows a hot-water supply system. (To make it more vivid, consider this as the supply system for the boss's morning shower.) Obviously, this system needs a temperature control system, and thus at some point we need to install a sensor to measure the controlled variable temperature. Where should this sensor be located? It is tempting to install this sensor at the point where the water will be used: at the shower. This is where the water will be used, where its temperature is important, and where all our best instincts lead us to install the sensor—and that is the wrong answer.

The sensor should be installed where the water leaves the hot-water tank itself. This provides much, much better control of temperature. If we were to install the sensor at the shower, then water would have to flow the distance from the tank to the shower, and the elapsed time would depend on the velocity of the water flowing through the pipe. The time required—a dead time—would be equal to distance divided by velocity. This type of dead time is usually called *transportation lag* for obvious reasons. With the sensor located at the shower, this dead time is inserted within the control loop. With the sensor at the tank exit, the dead time is outside of the control loop, and thus the loop will function far more quickly and effectively.

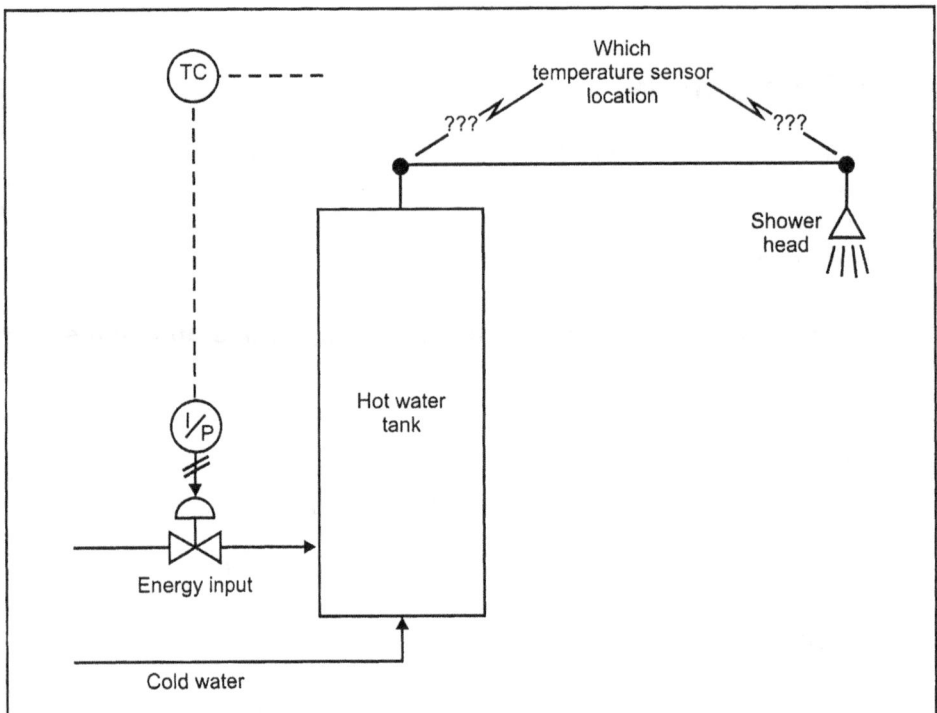

Figure 13-1. Hot Water Supply System

In the design and layout of control loops, sensors are often located at the wrong spot--though always with good motives. As a result, difficult control problems are often unnecessarily created. Sometimes equivalent problems are created when a sensor's location is dictated by ease of future maintenance, for example, rather than by considering whether or not it creates control problems.

The foregoing discussion is not meant to imply that a dead time is always avoidable. Quite the opposite is true. There are many cases where a dead time is inherent in the nature of the process itself; three common examples are shown in Fig. 13-2. In each case, there is a situation in which the material within the process must flow through equipment, and there is an inherent distance-velocity lag or dead time within the process equipment.

Another very common type of dead time is encountered in analytical equipment such as chromatographs or many, but not all, viscometers. A dead time is created by the time it takes these devices to operate and measure the value of the variable they are trying to determine. These sample times are, of course, dead times, but in a control situation they are located in the "feedback" part of the loop.

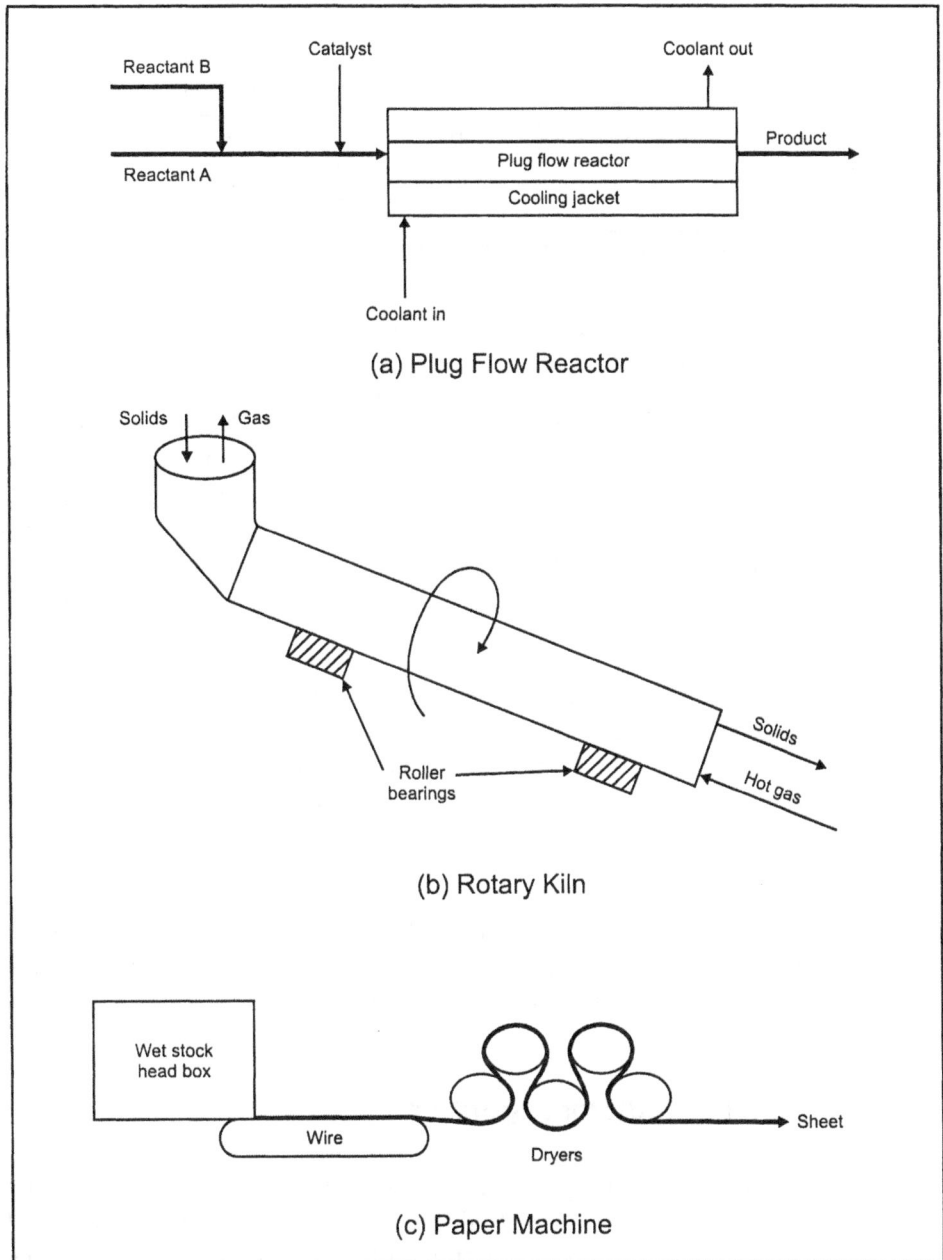

(a) Plug Flow Reactor

(b) Rotary Kiln

(c) Paper Machine

Figure 13-2. Typical Dead Times in Process Units

There are also other types of situations that are either dead times in a physical sense or that create the equivalent of a dead time insofar as a controller is concerned. The mathematical form of a dead time is shown in Fig. 13-3, in which we see the graphical picture of dead time θ. You can easily see that during the dead time q there is no output (in response to the input), and thus no output information is available to initiate control or

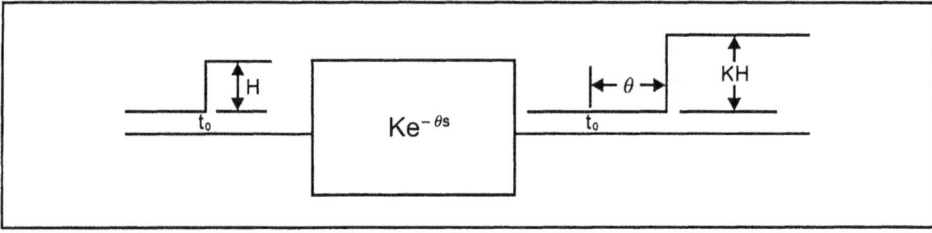

Figure 13-3. Mathematical Form of Dead Time

corrective action. This illustrates why dead time is the "most difficult element to control."

Representing dead time mathematically is also a bit awkward since conventional mathematics is formulated in terms of continuous variables, and a dead time implies a discontinuity. The notation used in Fig. 13-3 is the Laplace notation, in which a dead time θ is shown as $e^{-\theta s}$, where s is the Laplace variable. (This is the notation used throughout this unit because it is much more suitable than time domain notation when discontinuities are involved.)

One other characteristic of a dead time is also worth noting at this point. Because of the "wait" between the input and output of a dead time, there is "memory" involved, and analog hardware is thus insufficient. Digital hardware is always necessary to handle the control calculations in a reasonable manner.

Since dead time causes severe deterioration in control loop performance, there is real incentive to develop advanced control algorithms that can be used to control processes containing large dead times. In this unit, we will review several of these algorithms. Inevitably, they demand mathematical skills.

13-2. The Smith Predictor Algorithm

The Smith Predictor algorithm, developed by the late O. J. M. Smith (Ref. 9), is the best known of all the dead time compensation algorithms. To demonstrate it, we will assume a process that can be represented as a first-order lag plus a dead time shown as the following:

$$\frac{K}{1 + \tau s} \cdot e^{-\theta s} \tag{13-1}$$

In this equation, K is process gain, τ is process time constant, θ is dead time, and s is the Laplace variable. The block diagram for this equation is as shown in Fig. 13-4. As shown in Fig. 13-4, it is convenient to split the representation of the process into two parts, a first-order lag portion and a dead time portion. When this is done, it defines a fictitious variable B as shown in Fig. 13-5, which can then be fed back to the controller, also shown in Fig. 13-5. The dead time is now outside the loop, and the controlled variable C will do whatever B does but simply delayed by θ. There is, of course, no delay in the feedback of B. This schematic arrangement cannot be implemented, however, because B is fictitious and cannot be measured.

We can now develop a mathematical model of our process and use the manipulated variable M as an input to the model, as shown in Fig. 13-6. If there are no load disturbance inputs and if the model is perfect, then the model error in Fig. 13-6 is $E_m = 0$. We also see that we have a value of B_m to use as a feedback signal, as shown in Fig. 13-7. The scheme in Fig. 13-7 works well as long as our model is perfect and as long as no disturbances enter our loop. A second feedback loop can be added, however, as shown in Fig. 13-8, and this is the Smith Predictor control strategy. The strategy is usually implemented in an equivalent form, shown in Fig. 13-9, which also shows the implementation of digital computer control in which a sample on the loop operates every T units of time. The controller for the loop $G_c(s)$ can be a conventional PI feedback controller.

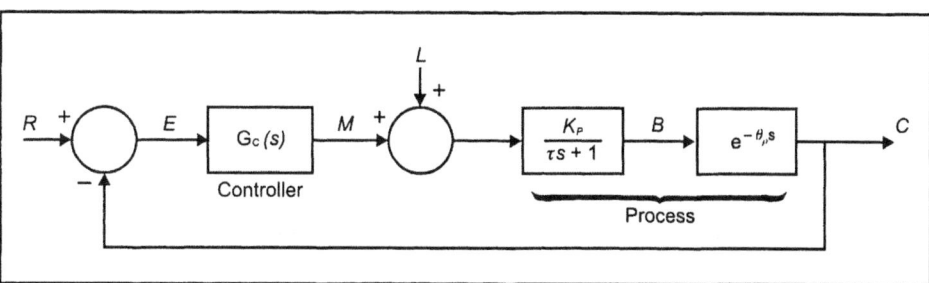

Figure 13-4. Feedback Control Loop with Process Dead Time

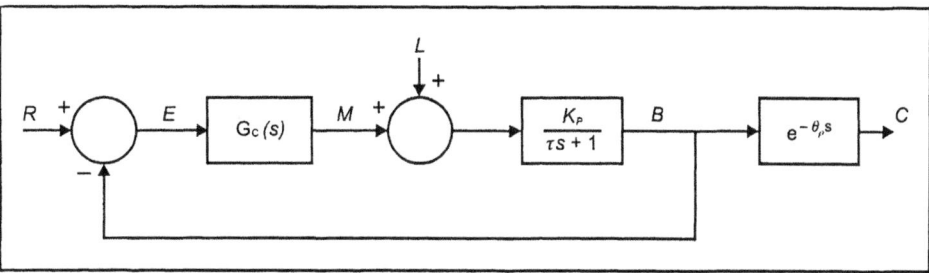

Figure 13-5. Desired Feedback Loop Configuration

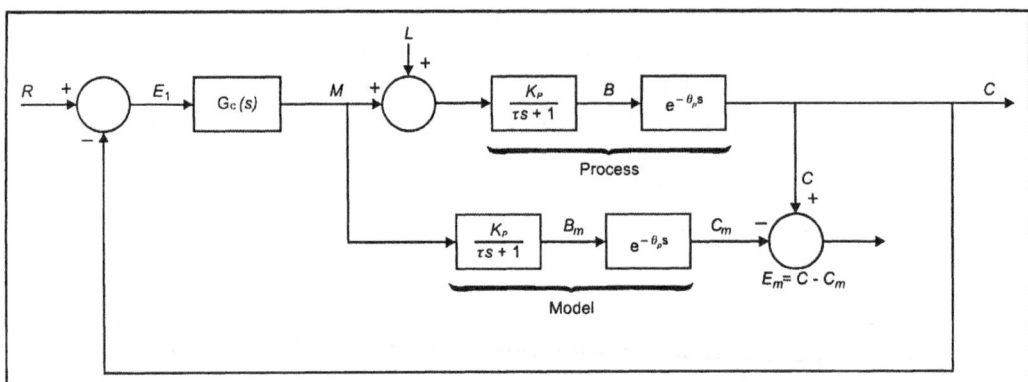

Figure 13-6. Using the Manipulated Variable as Model Input

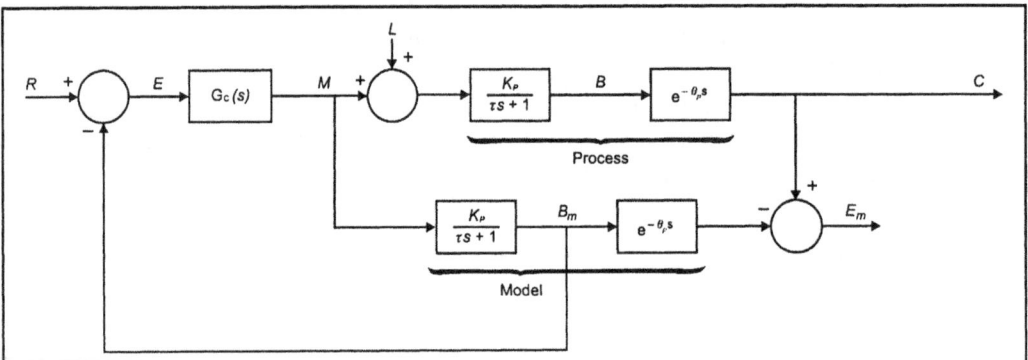

Figure 13-7. Using B_m for Feedback

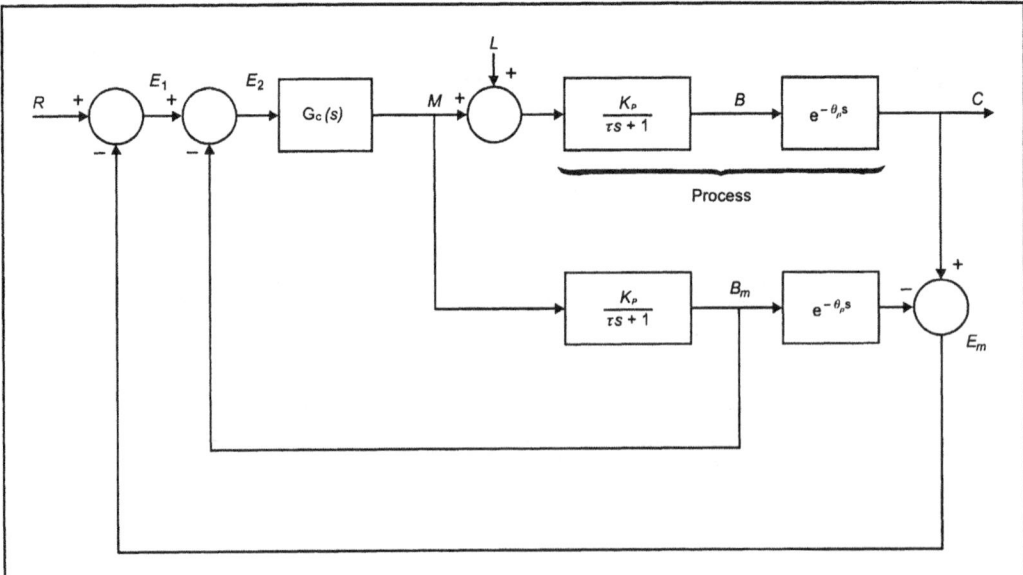

Figure 13-8. Smith Predictor Control Strategy

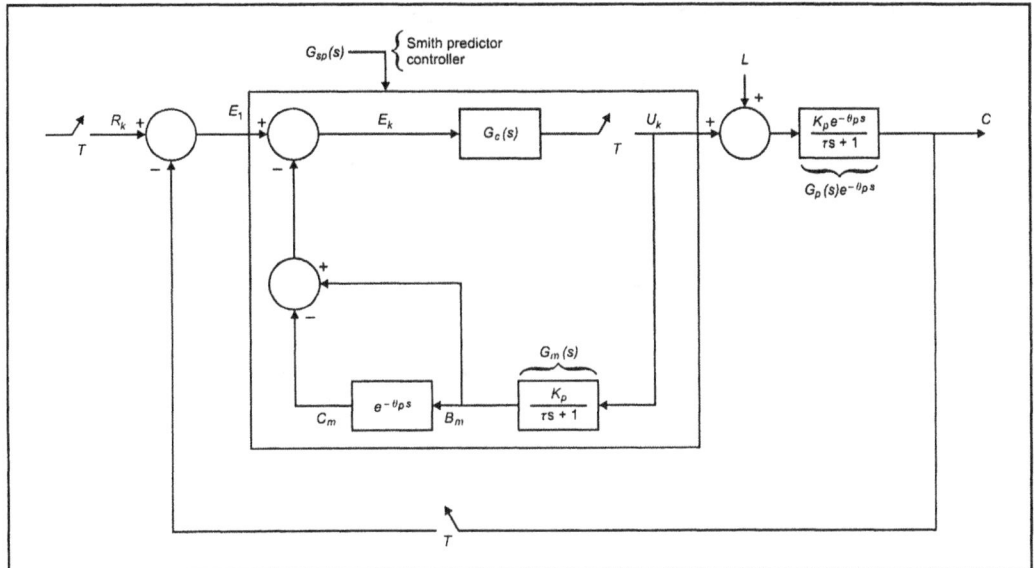

Figure 13-9. Computer Control Implementation of Smith Predictor

13-3. A Smith Predictor Application

Meyer et al. (Ref. 7) list several literature references discussing applications of the Smith Predictor. They also review the application results for a pilot distillation column similar to the one shown in Fig. 13-10, with a process described mathematically as follows:

$$X(s) = \frac{e^{-60s}}{1 + 1002s}R(s) + \frac{0.167e^{-486s}}{1 + 895s}F(s) \tag{13-2}$$

In this equation X is the top product composition; R is reflux; F is feed rate; and dead time and time constants are given in seconds. (Note that while the process does not have an obvious *transportation delay*, it is actually best represented by a mathematical model that includes dead time.) There is clear improvement in control because of the Smith Predictor.

A typical characteristic of the application of the Smith Predictor is that results are significantly affected by modeling errors. If the mathematical description of the process changes during operation, it makes it difficult or impossible to use the Smith Predictor.

A disadvantage of the Smith Predictor is that, although it requires a model of the process, it does not use that model to design or tune the feedback controller. Thus, it ends up with too many controller-tuning parameters: the model parameters plus the controller-tuning parameters. Because there are too many parameters to adjust, there is no convenient way to adjust the closed-loop response when the model does not fit the process properly. Given the nonlinear nature of process dynamics, any technique that depends too heavily on exact process modeling is doomed to fail.

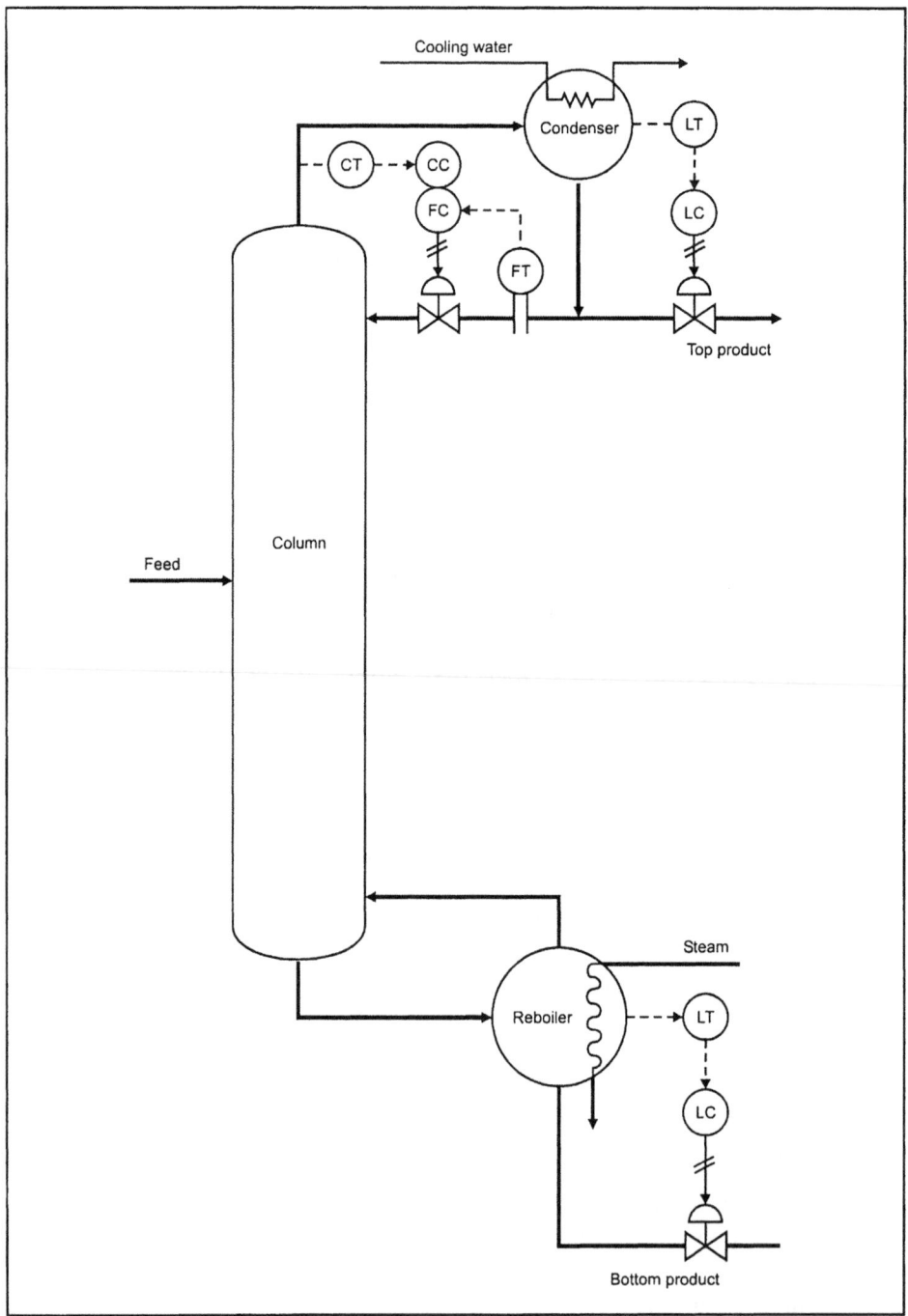

Figure 13-10. Pilot Distillation Column

13-4. The Moore Analytic Predictor

Moore (Ref. 8) has proposed an alternative to the Smith Predictor in which the process model is used to "predict" the future value of the controlled variable itself. This predicted value itself is then fed back and used as a controller input, as shown in Fig. 13-11. Moore has shown that digital computer control of a loop that has a sample time of T and a dead time q can be represented with an equivalent model dead time θ_m, where

$$\theta_m = \theta + \tfrac{1}{2} T \qquad (13\text{-}3)$$

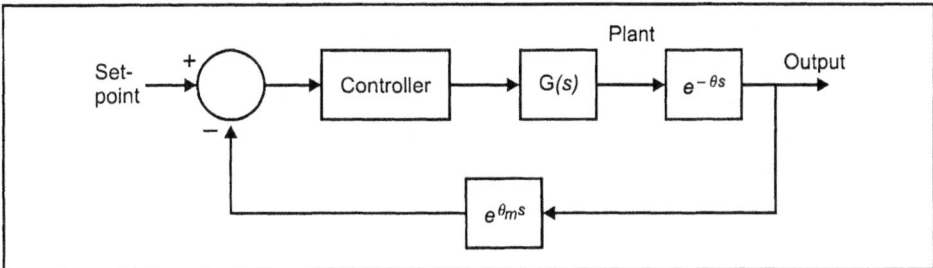

Figure 13-11. Moore's Direct Analytic Predictor

The analytic predictor then predicts the value of the controlled variable at θ_m time units in the future based on current inputs. Using the nomenclature of Fig. 13-11, the predictor uses the current value of the controlled variable c_0 and the values of the controller output $u(t)$ for the *post* θ_m units of time to estimate the process output at θ_m units of time in the *future*. If the process $G(s)$ is a first-order lag, then

$$G(s) = \frac{K}{1 + \tau s} \qquad (13\text{-}4)$$

The predictor solves the following equation:

$$\tau \frac{dc(t)}{dt} + c(t) = Ku(t) \qquad (13\text{-}5)$$

where the following initial condition applies:

$$c(t_0) = c_0 \qquad (13\text{-}6)$$

where t_0 is the current time and c_0 is the current output. The analytic solution is as follows:

$$c(t_0 + \theta_m) = c_0 e^{-t_0/\tau} + K \int_{t_0}^{(t_0 + \theta_m)} u(\lambda - \theta_m)e^{-(t_0 - \lambda)/\tau} d\lambda \qquad (13\text{-}7)$$

This integral can be solved numerically, or the original differential equation could be solved from time t_o until time $(t_o + \theta_m)$. A difference equation could also be developed. Moore's controller is PI, but the controller integral term is based on model error instead of control error. (This modification could also be incorporated into a Smith Predictor.)

13-5. The Dahlin Algorithm

The direct synthesis controller algorithm, also known as the Dahlin algorithm (Ref. 4), is a tailor-made digital controller algorithm. It is designed to provide a closed-loop response to set point changes that have no overshoot or oscillation if the loop has a fixed-time constant, dead time, or gain. The controller algorithm contains a complementary model of the process so as to provide a desired response to step set point changes. The desired closed-loop response is described by a selected dead time and time constant. This selected dead time must be close to the actual dead time of the loop to avoid instability.

For a loop with a pure dead time--that is, no time constant--the selected dead time must satisfy the following inequality:

$$0.5\,\theta_o < \theta_c < 1.25\,\theta_o \qquad\qquad (13\text{-}8)$$

where θ_o is open-loop response dead time, and θ_c is the dead time "selected" for the closed-loop response.

The range of selectable dead times becomes larger for loops that have a time constant. The larger the time constant, the wider the range. A difference equation for a direct synthesis controller for a loop with a first-order lag plus dead time self-regulating process is as follows:

$$O_n = B\,O_{n-1} + (1 - B)\,O_{n-N-1} + K\,(E_n - A\,E_{n-1}) \qquad (13\text{-}9)$$

where:

$$A = e^{-T/\theta_o} \qquad\qquad (13\text{-}10)$$

$$B = e^{-T/\theta_c} \qquad\qquad (13\text{-}11)$$

and where:

O = controller output
E = error (set point – measurement)
θ_o = time constant of open-loop response
θ_c = time constant selected for closed-loop response

$$T \quad = \quad \text{sample time of digital controller}$$
$$n \quad = \quad \text{subscript denoting present value sampled}$$
$$n\text{-}1 \quad = \quad \text{subscript denoting value from one sample period ago}$$
$$n\text{-}N \quad = \quad \text{subscript denoting value from N sample periods ago}$$
$$K \quad = \quad \text{gain factor}$$

The direct synthesis controller is identical to the Smith Predictor if closed-loop dead time is set equal to the open-loop dead time and the closed-loop time constant is set equal to the open-loop time constant. The Dahlin control algorithm is used extensively to control processes that have long dead times, such as paper machines. Corripio (Ref. 2) gives a good treatment of the use of the algorithm for dead time control.

13-6. Summary

In this unit, we have reviewed several approaches to dead time control. Additional special approaches and mathematical algorithms are available in the scientific literature, and the number increases each year. It never gets easy to do dead time control, however.

Some general observations concerning dead time control are in order here. First, whenever it is possible to avoid dead time in a control loop, it is wise to do so. Second, if you undertake dead time control, it will involve digital control hardware and some relatively advanced mathematics. Third, when these dead time algorithms are implemented, they rely on process models, and if modeling errors are present they can render the control ineffective. Model parameters need to be known within about 10 to 20 percent of actual values.

By now, you too are probably ready to agree that dead time is "the most difficult element to control."

REFERENCES

1.   K. J. Astrom and B. Wittenmark *Adaptive Control* (Addison Wesley, 1989).

2.   Armando B. Corripio, *Tuning of Industrial Control Systems* (ISA, 1990).

3.   P. B. Desphande and R. H. Ash, *Computer Process Control*, 2d ed. (ISA, 1989).

4.   E. B. Dahlin, "Designing and Tuning Digital Controllers," *Instruments and Control Systems*, vol. 41, no. 6 (June 1968), p. 77.

5.   Chang C. Hang, K. W. Lim, and B. W. Chong, "A Dual-Rate Adaptive Digital Smith Predictor," *Automatica* (January 1989), pp. 1-6.

6.   Chang C. Hang, T. H. Lee, and W. K. Ho, *Adaptive Control* (ISA, 1993).

7.   C. Meyer et al., "An Experimental Application of Time Delay Compensation Techniques to Distillation Column Control," *I &EC Process Design Development*, vol. 17, no. 1 (January 1978).

8.   C. F. Moore, "Improved Algorithms for Direct Digital Control, " *Instruments and Control Systems*, vol. 43, no. 1 (January 1970), pp.70-74.

9.   O. T. M. Smith, "Close Control of Loops with Dead Time," *Chemical Engineering Progress*, vol. 53, no. 5 (May 1957), pp. 217-19.

10.   E. F. Vogel and T. F. Edgar, "A New Dead-time Compensator for Digital Control," *Proc. ISA Annual Conf.* (ISA, 1980), pp. 29-46.

EXERCISES

13-1.   *Consider the following system, in which diesel oil is injected into fuel oil to control product viscosity:*

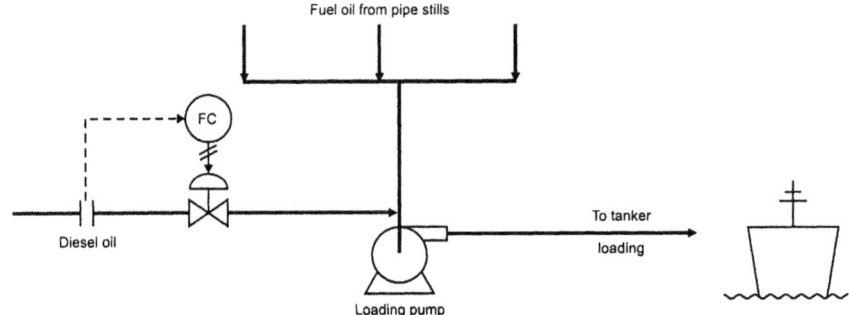

Where should a viscosimeter be located to measure viscosity and then, through cascade control, be used to provide a set point to the FC that controls diesel oil injection?

13-2.   *Discuss the dead time inherent in a steel rolling mill or in a paper machine.*

13-3.   *Compare dead time compensators as presented in this unit with conventional feedback control.*

13-4.   *In all the algorithms presented in this unit, the process model was assumed to be a first-order lag plus dead time. Can or should higher-order systems be considered when developing dead time compensation algorithms?*

13-5.   *Assume your digital control hardware vendor gives you a software program that implements an advanced dead time compensator--but you must supply the gain, the time constant, and the dead time for the process. How could you measure these experimentally?*

Unit 14:
Nonlinear Compensation and Adaptive Control

UNIT 14

Nonlinear Compensation and Adaptive Control

Virtually all process control theory is taught in terms of linear systems, while all real applications, to some extent, have nonlinear characteristics. There are two basic approaches for dealing with this contradiction: either compensate for the nonlinearities or adapt to them. We will discuss both in this unit.

Learning Objectives — When you have completed this unit, you should:

A. understand the character and implications of nonlinearities

B. appreciate the use and usefulness of compensation for nonlinearities

C. understand adaptive tuning

D. appreciate general adaptive control concepts

14-1. Nonlinearities

Linear is good; nonlinear is bad! Why? What does linearity mean anyway? Math professors talk about linearity, but does it affect anything in real life? When something is said to be linear, we think of a line--a straight line. We, therefore, think that something said to be linear can be extrapolated. Linear differential equations have general solution techniques, but for nonlinear differential equations we can only talk of specific solution techniques for specific equations under specific conditions. But what about control systems?

A control system is linear when it is described by linear mathematics. This implies that in certain senses its behavior is predictable. If we change an input by x, then allow the system to reach steady state, then change the input by an additional x, the new output will be exactly the same as, and in addition to, the output to the first x input. If we reduce the input by x, the output will be the exact mirror image of the output caused by an increase of x in the input. The output is a predictable linear extrapolation–both in extent and shape–of earlier outputs for different amounts of x.

When a control system is nonlinear, its response to a positive change of x is different from the mirror image of a response due to a negative change of x. The response and behavior of the system changes in time and changes as the load on the system changes. It needs to be retuned regularly, while a linear system may be tuned once and stay tuned forever more. Linear systems behave in a predictable way and may be operated more tightly.

Linear is good; nonlinear is bad.

In general, linear systems have the following characteristics:

1.   All dependent variables are continuous; that is, they give a continuous response over the range of values exhibited by the system.

2.   All coefficients of the differential equations that characterize the system are constants or functions of time only.

3.   When several inputs are applied at the same or different locations at the same or a different time, the total response is the sum of the individual responses (superposition).

4.   The magnitude of output is directly proportional to the magnitude of the input.

Nonlinear systems lack one or more of these characteristics. Much can be done when designing control systems to cause them to be linear. And when it is inevitable that nonlinearities are present--due either to inherent characteristics or to changes in parameters in time or under varying loads--we can sometimes install compensation so that the overall result is linear.

In a control sense, it is also possible to design a system that adapts itself to varying parameters. Adaptive control is an imprecise term, but most often it is used to deal with nonlinearities that cause process conditions (parameters) to change. The control strategy or controller characteristics are adapted (changed) during actual operations so as to maintain good (best?) control of these changing conditions. Most adaptive control systems have some scheme of process identification, whose results are then used to adjust one or more parameters in the control algorithm. As an example, process gain is measured and controller gain is adjusted so that the product of the two is always constant.

In general, it is usually better to avoid or compensate for nonlinearities. If that cannot be done, adaptive control can then be considered.

14-2. Valve Characteristics

Control valves are a good place to begin a consideration of nonlinearities. The flow characteristics of control valves generally fall into three broad categories, as was shown in Fig. 7-2. Refer to Fig. 7-2 now. These characteristic curves can be misleading. Superficially, it appears that classes A and C are nonlinear, while B is linear--but the problem is more complex. To understand control valve behavior, it is most important that you gain a specific understanding of the dynamic characteristics of a control valve installed as a part of a total piping network. The concepts involved significantly affect the valves' operating characteristics and the total performance of the feedback control system.

At this point, we urge you to go back and restudy all of Section 7-7, "Control Valve Dynamic Performance," in detail. The result of all this is that a linear control valve may not behave as a linear device, while an equal-percentage valve may operate over a range in which its behavior is effectively linear. You must consider the family of valve curves, and you must make an analysis of system pressure losses in order to understand a specific valve's behavior. Only then can you intelligently design the control system.

14-3. Process Characteristics

Is the process itself linear or nonlinear? Consider the heat exchanger shown in Fig. 14-1. At steady state the energy balance is as follows:

$$WC_p(T_o - T_i) = F\Delta H \tag{14-1}$$

where

W = heated liquid flow rate
C_p = heat capacity of liquid
T_o = outlet temperature
T_i = inlet temperature
F = steam flow rate
ΔH = steam enthalpy change

Solving for F:

$$F = \frac{WC_p(T_o - T_i)}{\Delta H} \tag{14-2}$$

Equation (14-2) is nonlinear if we assume that W varies and since we know that T_o varies. But if we have a steady state involving one value of W, consider what happens if we double W. Using these two situations as example base cases, for each one consider what happens when T_o is below set point by 5°F. The same error exists, and the same signal goes to the control valve. But in the case where W is doubled, we need twice the increase in steam. We really need an equal-percentage increase in steam; we really need true equal-percentage trim and behavior from the control valve. When this happens, the total combination of the nonlinear exchanger and the nonlinear equal-percentage valve is linear.

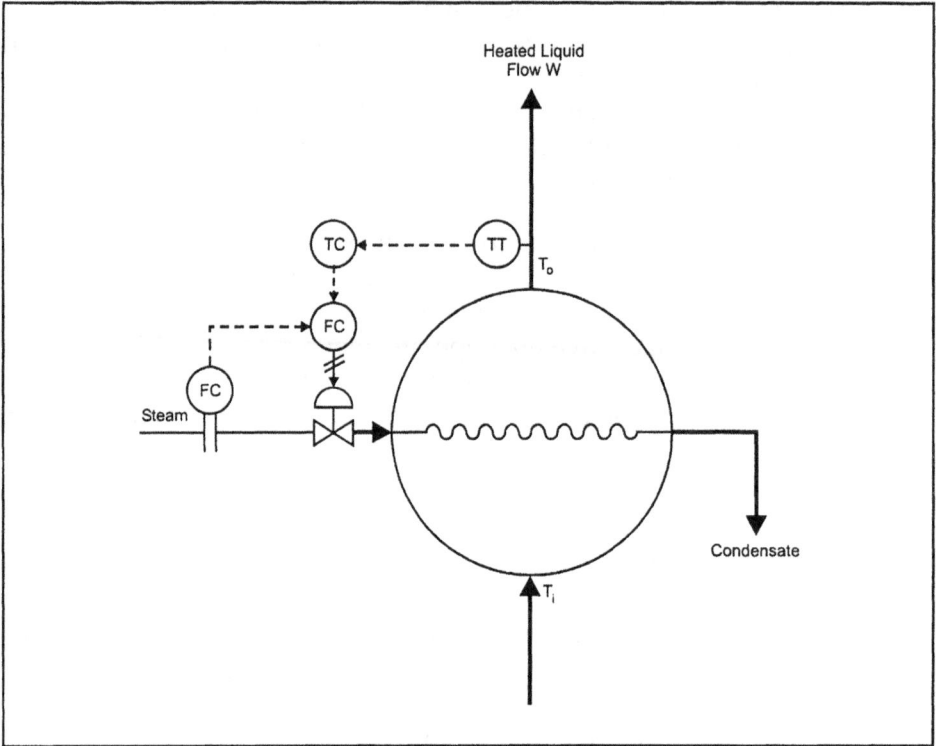

Figure 14-1. Heat Exchanger System

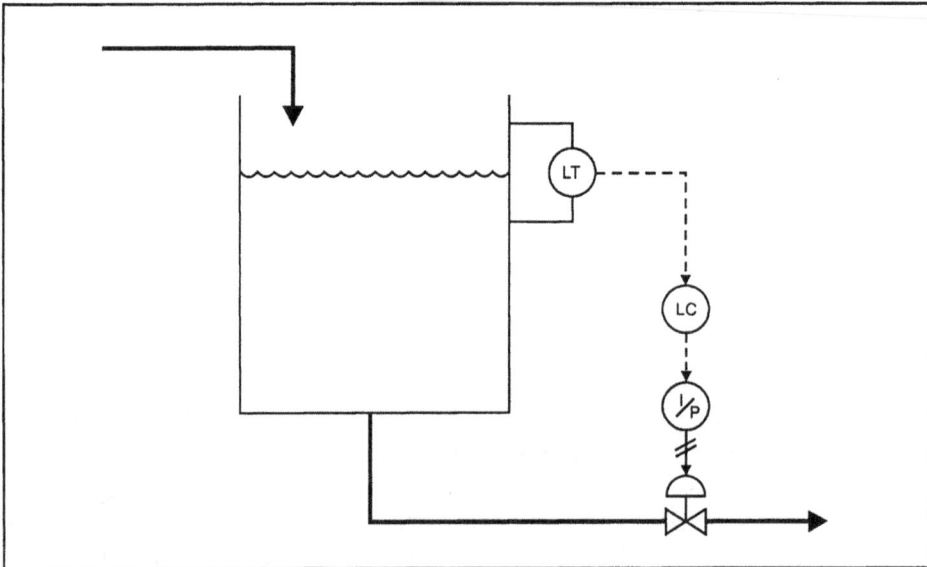

Figure 14-2. Liquid Level System

Now consider the liquid-level control system of Fig. 14-2. This process (tank system) is linear. When the level is low by 1 inch, for example, we get the same signal to the valve no matter where the set point is. That is good because we need the same change in outlet flow since the tank sides are straight. The system is linear, and for the combined system to be linear we need linear trim and behavior in the control valve.

The process itself may have nonlinear characteristics that are inherent and cannot be designed around. Consider a pH process, for example. A titration curve for a strong acid and base is shown in Fig. 14-3. Even where the curve appears to be a vertical straight line (between 2 and 12 pH), however, a blowup of this curve in that region yields additional S-shaped regions (nonlinearities). These major nonlinearities make a pH process very, very difficult to control and virtually create pH control as a subject all its own.

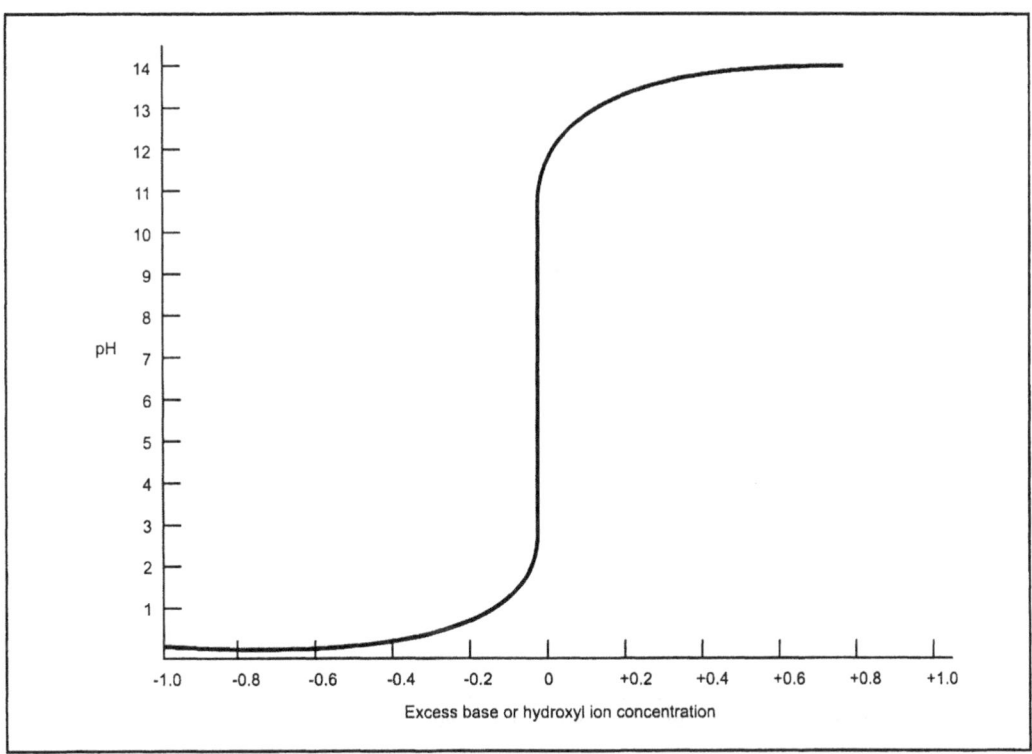

Figure 14-3. Titration Curve

In Section 8-4, we reviewed a very special type of nonlinearity, dead time, and saw that it created the need for special control considerations. Sensors themselves can also create special nonlinear problems. For example, differential-pressure flow sensors transmit a signal that is proportional to flow squared, and in some cases it is desirable to use a square root extractor in connection with such transmitters. Many transmitters— chromatographs are a good example—operate in definite cycles, and their output is not continuous but, instead, consists of a series of discrete data points.

Some processes even exhibit a nonlinearity of an "inverse" nature, such as is shown in Fig. 14-4. This is the response of a boiler drum to an increase in feedwater flow. What happens is that cooler water flows into the drum, some steam bubbles are condensed, the apparent level falls, and then the level rises as more new water enters and new steam bubbles are formed.

The examples could go on and on. Nonlinearities can sometimes be avoided by good analytical design or by wise equipment selection and specification or by the use of special compensators. But sometimes they are unavoidable and must be dealt with by using special control techniques.

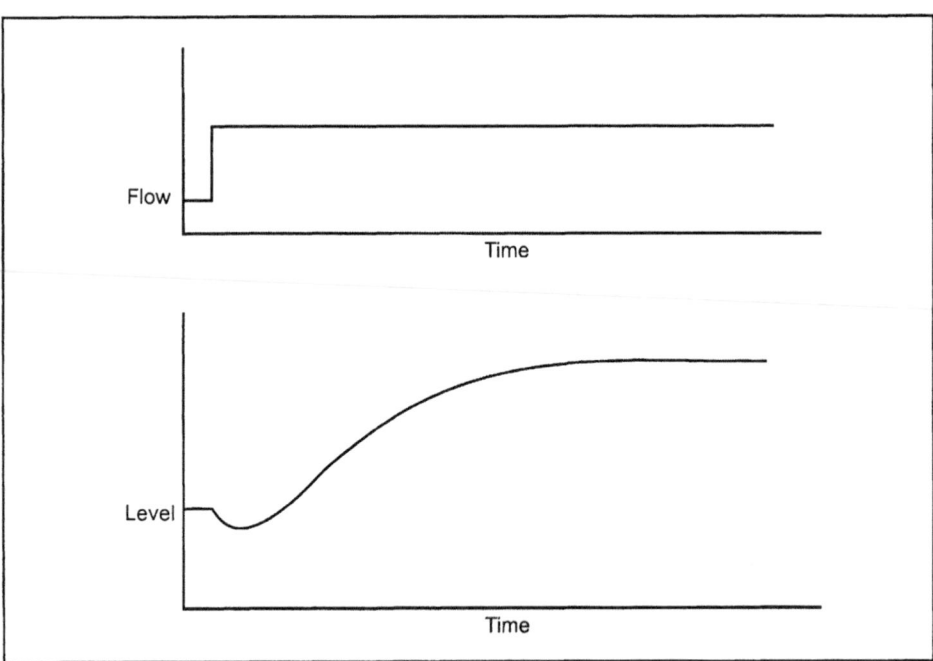

Figure 14-4. Inverse Response of Boiler Drum Level

14-4. Adaptive Control

Suppose that we compensate for nonlinearities by introducing suitable nonlinear functions, at the controller input or output, for example, or we select a particular valve trim to offset process characteristics. Neither of these is thought of as adaptation because these functions (compensation) are fixed. Adaptive control is based on the development of a compensating adjustment through knowledge of a disturbing factor or on the basis of loop response itself. A system that adapts itself on the basis of a measurement of a disturbing factor is referred to as a *programmed system*. A system that uses a measurement of its own performance is termed *self-adaptive*.

Let's tackle programmed adaptation first. In those processes where a measurable process variable produces a predictable effect on the gain of the control loop, we can program compensation into the control system. Easily the most common example of this in process control is the variation in dynamic gain with flow in longitudinal equipment where no back-mixing is evident. It is common in heat exchangers (recall Fig. 14-1), but it is a particularly severe problem in once-through boilers. Fig. 14-5 shows the response of steam temperature to a step change in firing rate. At 50 percent flow, the steady state gain is twice as high as a 100 percent flow because only half as much water is available to absorb the same increase in heat input. The dead time and time constants are also twice as great when the system is at 50 percent flow.

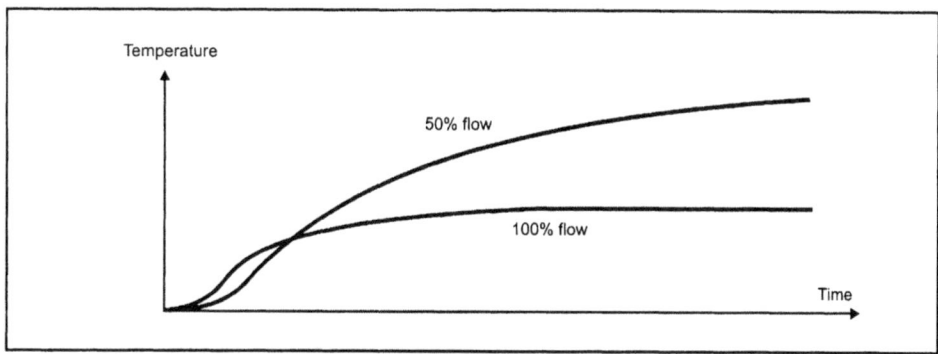

Figure 14-5. Boiler Response to Firing Rate Change

The controller algorithm gains can be programmed to vary with the fraction of full-scale flow. Proportional band should vary inversely with flow. Since the process dead time and time constants vary inversely with flow, the reset time T_i and the derivative time T_d should also vary inversely with flow. The flow-adapted three-mode algorithm becomes the following:

$$m = \frac{100w}{PB}\left(e + \frac{w}{T_i}\int e\,dt + \frac{T_d}{w}\frac{de}{dt}\right) \qquad (14\text{-}3)$$

The principles involved in such programmed adaptive control can be extended to other process situations as well.

On the other hand, self-adaptive systems must sense variations in plant behavior. This is referred to as *identification*, and it is normally a partial measurement because total identification is either too complex or impossible. After identification, the adaptive device or system must be able to adjust some controller parameter or function. This is referred to as actuation and is illustrated in the schematic diagram of Fig. 14-6. The identification, actuator, and control functions of a system are limited only by the skill and ingenuity of the system designer. They may be simple or complex, linear or nonlinear, continuous or discontinuous.

In a real sense, programmed adaptive control is feedforward in nature, while self-adaptive is feedback.

One final note: The definition of adaptive control we have used here is broad compared to some usage you may encounter. Some authors restrict the use of the term *adaptive control* to those systems that measure, and attempt to control, some index of performance. In such cases, systems like those described in the preceding paragraphs are called *nonlinear control systems*.

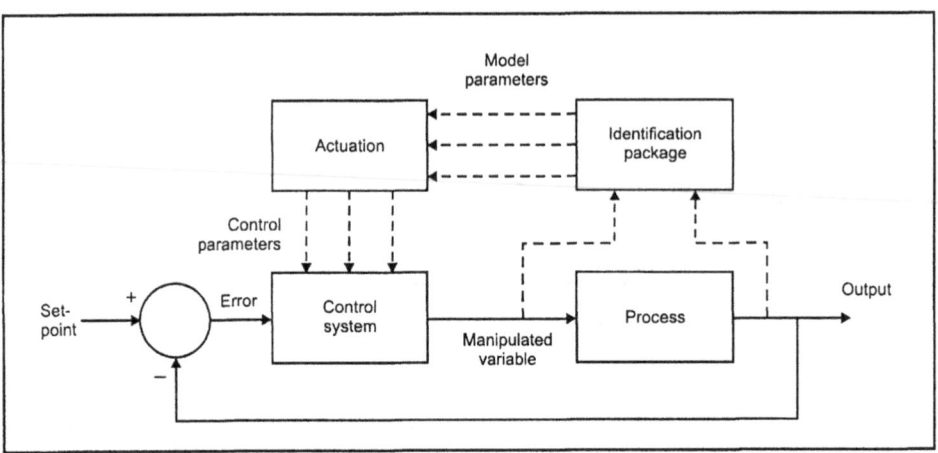

Figure 14-6. Self-Adaptive System

14-5. Adaptive Gain Control

Adaptive gain control is probably the single most commonly employed self-adaptive control technique. For example, all jet aircraft have adaptive gain control so the controls will always "feel" the same to the pilot whether the plane is at sea level or at 35,000 feet and whether the plane is empty of cargo or full. There are many different techniques for maintaining the product of all the gains around a loop as a constant.

Once in a while, it is possible to easily measure or infer the variable gain of a component and, if so, to adjust other gains to compensate for changes. For example, control valves often have nonlinear gains, as Fig. 14-7 illustrates. If such a curve is known, then the gain of the valve can be determined by measuring stem position. By neglecting valve dynamics, it is possible to change the gain in the control algorithm so as to keep the product of the controller gain and the valve gain constant.

A much different technique for adaptive gain control is that proposed by Marx (Ref. 6). Refer to Fig. 14-8 and assume that the objective is to maintain the product of the controller gain K_c and the process gain K at a constant value. Marx's technique is based on an analysis of the frequency characteristics of the error signal. As the loop's total gain KK_c increases, the transmission of high-frequency noise around the loop will also increase. This is illustrated in Fig. 14-9, which plots the frequency spectra of the error signal (produced by a step change in set point) for different values of loop gain KK_c. As loop gain KK_c increases, the amplitude at low values of w decreases and the amplitude at high values of w increases. The frequency servo of Fig. 14-8 uses a high-pass filter and a low-pass filter, the outputs of the two are compared within the servo and their relative values are used to change the controller gain K_c. This technique has been improved (Refs. 2 and 7) by adding high and low bandpass filters using weighting functions to produce the configuration illustrated in Fig. 14-10.

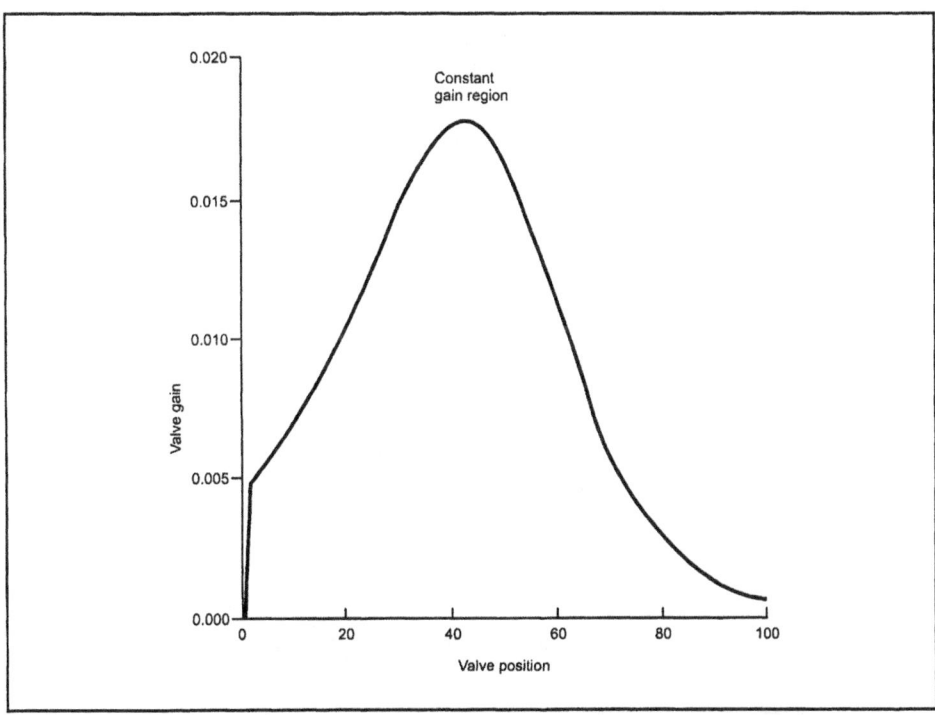

Figure 14-7. Installed Valve Gain of an Equal Percentage Inherent Characteristic

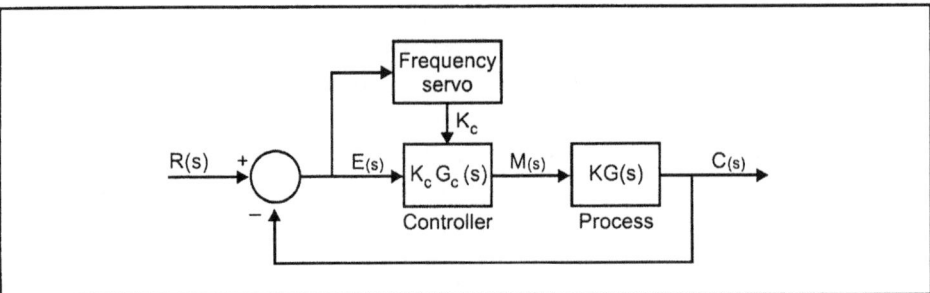

Figure 14-8. Marx Frequency Servo (Ref. 6)

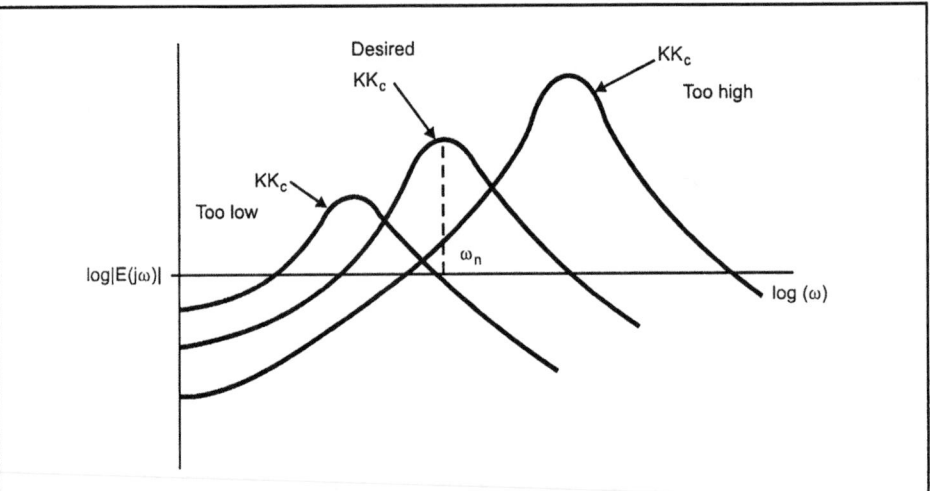

Figure 14-9. Error Spectra Various Values of Loop Gain

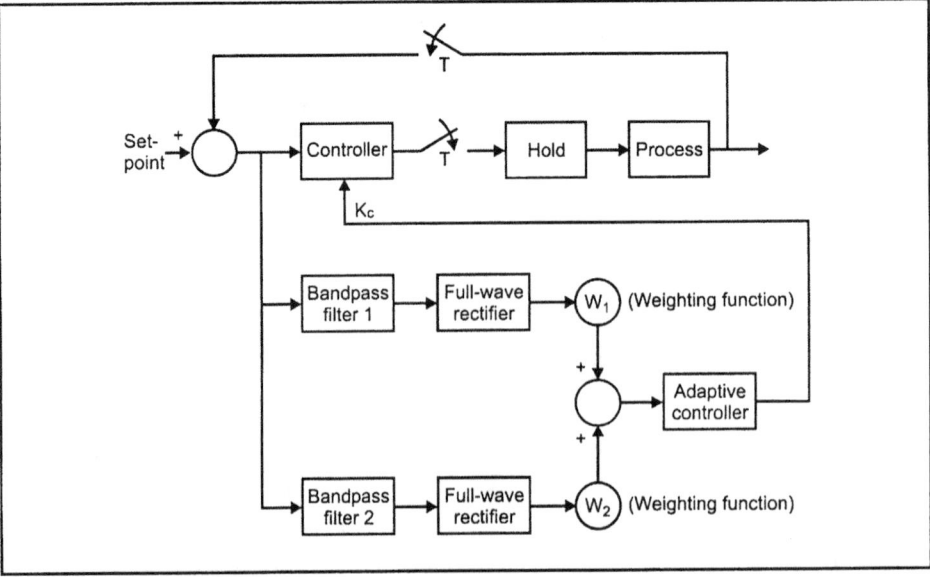

Figure 14-10. Bakke's Adaptive Gain Tuning

Other adaptive gain tuning techniques are available, but for illustrative purposes the ones shown in Figs. 14-12 through 14-14 are sufficient.

14-6. Three-Mode On-line Tuning

The concepts of adaptive gain tuning can be extended to adjust all the parameters in the controller algorithm. For example, as part of a computer control system, it is possible to have the computer periodically pulse the system, identify process characteristic coefficients, and then calculate parameters to tune the controller. This is illustrated in Fig. 14-11.

If you attempt to characterize the process as a first-order lag plus dead time, you would use the process reaction curve to determine the process parameters K, θ, and τ. You would then relate these to proportional gain, reset time, and derivative time by way of tuning maps.

One of the most successful approaches to PID adaptive tuning is based on a pattern-recognition approach developed by Bristol (Ref. 3). Bristol's article not only gives an outline of the technique but also provides an excellent discussion of the state of adaptive control technology and philosophy. The pattern-recognition technique is based on the interpretation of transient disturbances, in particular those that result from stepped set point or load changes.

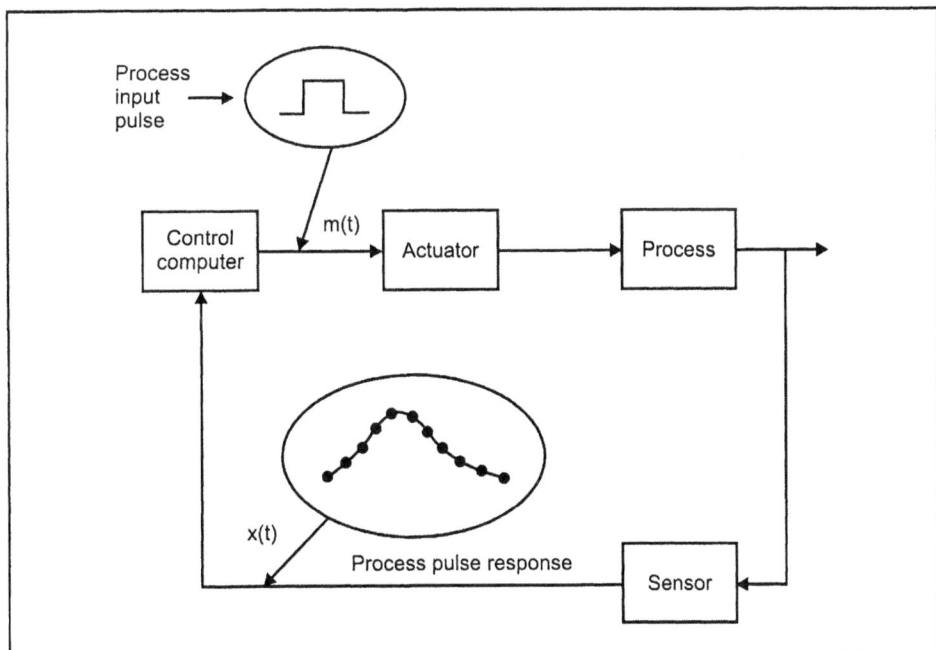

Figure 14-11. Pulse Testing a Process

The pattern-recognition design analyzes the shape of the time responses of the controlled variable to transient set point changes or load disturbances and compares the actual response characteristics to desired response characteristics. Differences between measured and desired responses generate response error signals that are used to adapt the parameters. Viewed in greater detail, as shown in Fig. 14-12, the analysis and adaptation to a set point change or load disturbance can be thought of in terms of five basic stages:

1. Recognition of a new disturbance with a peak error that is larger than a predefined noise threshold. (The transient behavior is in every case measured against the steady state that precedes the disturbance.)

2. During the process recovery, identifiying the T_L, the time to go from 75 percent of peak error to 25 percent of peak error. The times T_1, T_2, and T_3 in Fig. 14-12 are proportioned to T_L.

3. Integration of the controlled variable error response from T_1 and T_2 so as to compute S_1 and from T_2 to T_3 so as to compute S_2. Both S_1 and S_2 are further normalized by being divided by the original transient peak error and the corresponding integrating time. S_1 and S_2 are the pattern features used for adaptation.

4. Response errors are generated as the differences between S_1 and R_1 and between S_2 and R_2. Note that R_1 and R_2 are reference values that can be set to reflect any reasonable response, damped or resonant.

5. The two errors are used in a decoupling feedback to adjust the P and I settings of a standard PID. (When the derivative is set, it is done so by using a more complex algorithm, which we will not address here.)

The actual implementation is as shown in Fig. 14-13. This pattern-recognition system has been very, very successful and serves as a general-purpose adaptive controller to keep simple loops tightly tuned. Inherent in the approach is exploitation of the fact that within a given range of normal operations, loops can be treated as nearly linear and easily decoupled. Since the system continues to adapt as the operating state changes, this is normally acceptable.

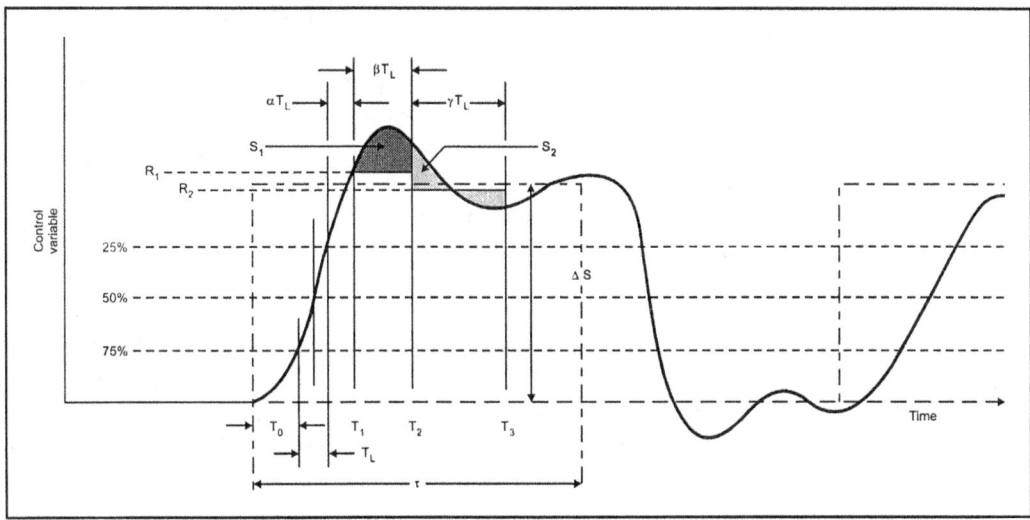

Figure 14-12. The Transient Response of the Process as Identified by the Adaptation Scheme (Ref. 3)

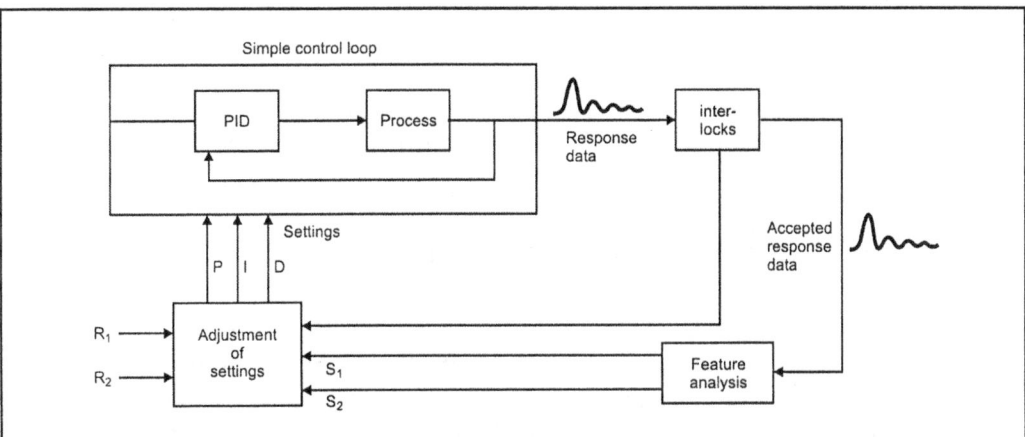

Figure 14-13. Design Structure

In time, perhaps adaptive tuning can be extended to other control situations in which there are few successful installations today. These include the following:

1. *Control of multivariable processes.* From a practical standpoint, true multivariable adaptation will be much more difficult to achieve because of the greater process complexity per process measurement and the correspondingly greater difficulty involved in identifying the process. It is possible, however, that single-loop adaptive controllers in multi-loop applications will satisfy most requirements.

2. *Control of nonlinear processes.* General nonlinear adaptation has not been achieved experimentally, even though some methods for doing so have been proposed. But many practitioners have proposed saving the settings developed by adaptation as a function of an operating point or batch recipe.

3. *Feedforward control.* Adaptation of feedforward control is more specialized and easier than adaptation of feedback loops.

4. *Control of processes with special dynamics.* Examples of this situation are processes that have true dead time, infinite right-half plane zeros ("shrink swell" or "reverse action" processes) in the transfer function, integrating processes, resonant processes, and unstable processes. Many of these processes present special technical problems under adaptation.

5. *Cascaded loops.* Cascaded loops are like any other single loop, with one exception. The inner loop, if properly designed, should be significantly faster than the outer loop. Thus, the control actions from the outer loop are too slow to be the basis of good adaptation.

Most control systems vendors are now supplying adaptive-tuning hardware and software, and their literature is a good source of current practice.

14-7. Model Reference Adaptive Control

As a last example of adaptive control, consider the so-called model-reference system shown in Fig. 14-14. In this type of system, a reference model is proposed, and the model's response is compared with the process's response. The difference between these two is an error e´(t), which is used to adjust or adapt the system's controller so as to minimize the error with respect to some figure of merit, such as the following:

$$\text{Figure of merit} = \int_0^t [e'(t)]^2 dt \qquad (14\text{-}4)$$

The adaptive behavior is superimposed on conventional feedback control.

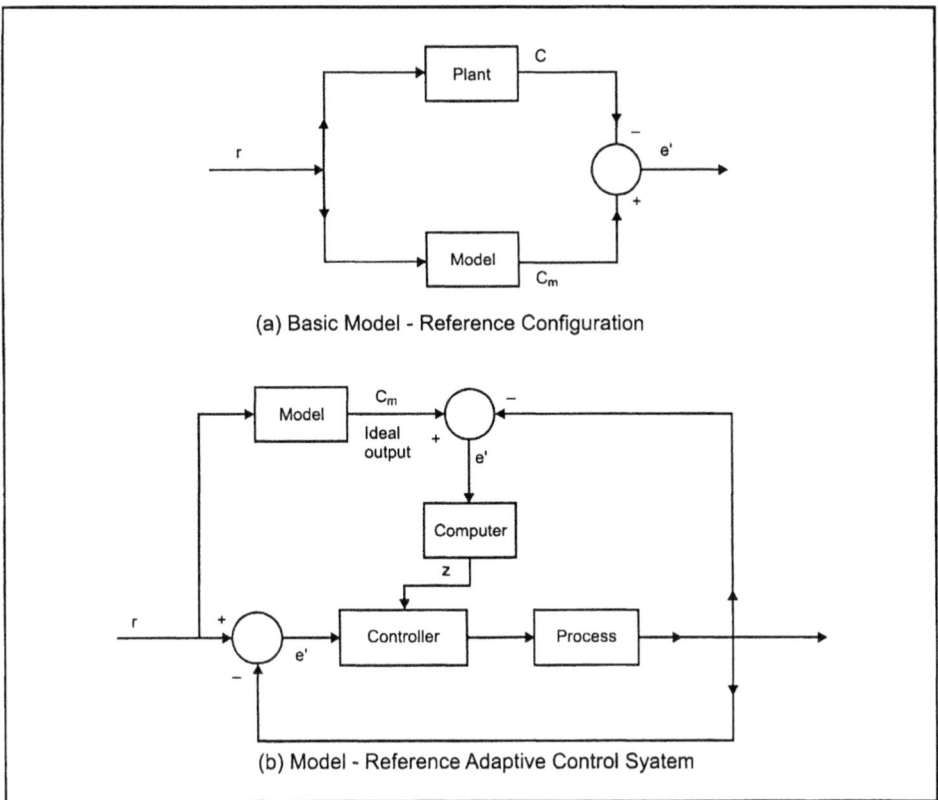

(a) Basic Model - Reference Configuration

(b) Model - Reference Adaptive Control Syatem

Figure 14-14. Model-Reference Adaptive Control

REFERENCES

1. K. J. Astrom and B. Wittenmark, *Adaptive Control* (Addison Wesley, 1989).

2. R. M. Bakke, "Adaptive Gain Tuning Applied to Process Control." Paper presented at 19th Annual Instrument Society of America Conference and Exhibit, Oct. 12-15, 1964, New York. (ISA reprint 3.2-1-64)

3. E. H. Bristol, "The Design of Industrially Useful Adaptive Controllers," *Advances in Instrumentation*, vol. 37, pt. 2 (ISA, 1982).

4. Armando B. Corripio, *Tuning of Industrial Control Systems* (ISA, 1990).

5. Chang C. Hang, T. H. Lee, and W. K. Ho, *Adaptive Control* (ISA, 1993).

6. M. F. Marx, "Recent Adaptive Control Work at the General Electric Company," *Proceedings of the Self-adaptive Flight Controls Symposium*, January 1959.

7. E. Mishkin and L. Braun, *Adaptive Control Systems* (McGraw-Hill, 1961), pp. 327-32.

EXERCISES

14-1. *Given Equation (14-1), what is the steady-state process gain? How does this vary with flow?*

14-2. *Given the following installed characteristic curve for a butterfly valve, how could you compensate for the nonlinear character of the valve?:*

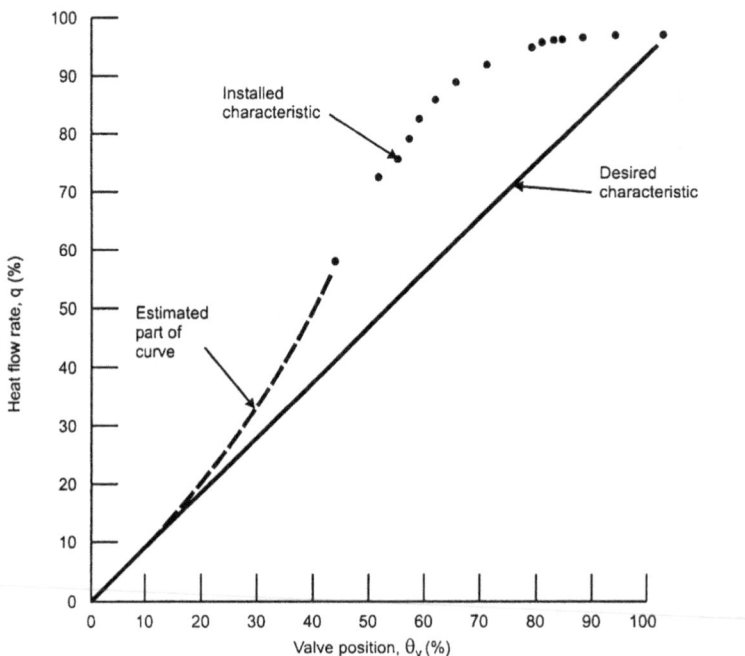

14-3. *How could you implement Equation (14-3) in practice?*

14-4. *Select one vendor's on-line adaptive-tuning system, and see what theoretical principles they use to adapt the controller algorithm.*

Unit 15:
Modern Control
System Architecture

UNIT 15

Modern Control System Architecture

This unit focuses on the components of individual control loops and how these components are assembled into large control systems. The structure of these larger systems is also illustrated, and some insight into their advantages and disadvantages is provided.

Learning Objectives — When you have completed this unit, you should:

A. be able to identify and discuss the basic hardware components of control loops

B. be able to identify and discuss the devices used to interconnect components together into control systems

C. be able to sketch and describe a direct digital control (DDC) system

D. be able to sketch and illustrate a supervisory control system

E. be able to sketch and illustrate a distributed control system (DCS)

F. be able to discuss the advantages and disadvantages of single-loop versus multi-loop control systems

15-1. Basic Components

The modern control system architecture has five basic components: sensors and transmission components, contact devices, final control elements, controllers, and display/recording components (Ref. 4). We will discuss each of these in turn in this section.

Sensors and Transmission

Temperature, flow, pressure, and level are the four major variables measured in process control systems. However, many, many other variables, such as pH, speed, humidity, position, composition, and so on, may also be measured. The sensors that make these measurements may be simple, or they may be very complex. They are connected from their field location to a display, record, and control location that may be nearby or remote. The transmitted signal may take many different forms, but today 4 to 20 mA is the most common standard. Most variables have their own unique twisted-pair wire connection, but signals may be multiplexed to reduce wiring costs.

Some sensors are inherently digital, and the trend is toward providing more and more sensors that have digital output capabilities. Also, many sensors are now being made "smart" with an internal microprocessor so better measurements are available to the control system. Digital sensor output can be converted back to analog for transmission on a 4 to 20 mA wire, but there is a loss of accuracy and repeatability. This increases the need for digital fieldbuses, and they are now starting to be installed.

Contact Devices

On-off signals are a part of virtually any control installation. They are needed to indicate process or equipment status, for example, valve position indicators, safety devices, motor starter condition, level switches, and so on. Such signals are used to indicate status, open and close valves, and start up and shut down equipment, among other purposes.

Final Control Elements

Control valves are most often the final control element in a loop. These valves usually have a pneumatic actuator to position the valve. Signals from the controller to the final control element are usually 4 to 20 mA, and these must be converted to an air pressure, usually 3 to 15 psig, to operate the actuator. Most valve positioners have this I/P capability. It is also possible to have valve positioners that accept a digital signal directly, and the trends are moving increasingly in this direction.

There are other types of final control elements besides control valves. Examples include variable-speed drives for pumps and motors, on-off devices, and others. Appropriate signal conversions are needed to convert the signal being transmitted from the controller to the form needed by the specific actuator.

Controllers

Analog controllers, either pneumatic or electronic, were the standard control devices for many years. Many of these analog controllers had digital interfaces that allowed them to be switched from manual to automatic. In virtually all cases, however, the tuning of the controller was done manually.

Most new controllers are microprocessor-based. Many are designed to handle a single loop, while others can handle multi-loop arrangements (see Section 15-7). They provide PID-type control, but much more advanced algorithms are readily available. Many provide for automatic tuning. Interfaces to other control system equipment are facilitated via digital interfaces, usually operating in conjunction with a bus (local area network or LAN).

Display/Recording Components

Older analog control systems are oriented toward using large panels to display and record data. Typically, these displays are located in a separate control room that also contains termination racks for field wiring and perhaps even a separate computer console for control-related purposes. Sometimes, the data display and recording equipment are shared among many loops. In older systems, each loop may have its own.

Most newer systems are digitally oriented and use video display units to display information. The graphics that are used to facilitate this display can be very impressive. Operators input data or commands to the system through a keyboard, light pen, or touch panel. Multiple video display units are often used. These digitally oriented operator stations can be centrally located, or they can be distributed throughout the plant.

15-2. System Components

The five basic system components are computers, bulk storage devices, software, field networks, and local area networks (LANs).

Computers

In the 1920s, a machine shop usually had a large, elevated motor at one end and a long driveshaft running down the length of the room. Individual machines used a belt drive for a power takeoff from the shaft. Today, each individual machine not only has its own motor, but virtually all machines have many smaller motors to provide power wherever it is needed.

As computers began coming into use in process control, there usually was one central machine that did everything. Today, as in the machine shop, wherever computational power is needed a computer of some sort is installed. Small, special-purpose computers are connected with larger machines over LANs. Programmable logic controllers (PLCs), first used in discrete manufacturing in the automobile industry in 1970, are also increasingly used today. They replaced hard-wired electromechanical relays and provided greater flexibility.

Now we have personal computers (PCs) that can be assigned special functions for process control.

Bulk Storage Devices

In addition to the normal storage needed by individual computers, there is also a need for bulk storage. This bulk storage is needed for the following typical uses:

- to archive plant operating records
- to store plant design data and flowsheets
- to maintain operating instructions, procedures, and recipes
- for operator training facilities and procedures
- for troubleshooting and/or design studies

The need for storage always grows; the cost drifts lower as new technologies become available. Most current storage is by way of electromechanical disk storage, but laser-optical systems are rapidly coming on the scene.

Software

System operating software is vendor supplied. It allows the user to easily implement a control strategy, but the software itself cannot be (and should not be) modified by the user. The vendor normally supplies a programming language that allows the user to write control applications. In addition, special software packages such as statistical process control, artificial intelligence, and optimization packages are usually available. Database software is also routinely available.

Field Networks

Field data transmission often is based on twisted-pair copper wires, using a 4–20 mA signal, and individual wire pairs are often used for each variable (Ref. 3). The power to operate the field device is often provided through this same pair. As you might expect, this structure results in the use of literally thousands of pairs of twisted wires. This wiring usually costs more than the control equipment itself. The cost is driven even higher by the need for explosion-proof or intrinsically safe installation and/or the need to prevent ground loops.

Because of the high costs of field wiring, the increasing load placed on transmission systems, and the need to transmit digital signals directly, a standard digital communication system is a clear necessity. *Fieldbus* is the term used to describe the digital replacement for the 4-to-20 mA DC as the method for communicating between the elements of a distributed process control system. Fieldbus needs to provide for two-way communication that allows full access to intelligence located in field devices.

The characteristics of the digital fieldbus are such that it must be very secure, and this is usually accomplished through redundancy. It must have a superior capacity to recover the signal from the typical electrical noise found in manufacturing environments. The fieldbus must also be fast, with response times from microseconds to milliseconds, and it must also be capable of powering field devices over long distances. There also is the need for electrical isolation and for intrinsic safety. Finally, there needs to be protocol support for the elements of a real-time operating system, which is needed to synchronize the time-critical functions in many devices that are only connected by a communications network.

Several proprietary fieldbuses have been on the market for some time, but there is an obvious need for standards. ISA started an effort in the late 1980s to develop such standards, and now after many years they are nearing reality. See American National Standards Institute standard ANS/S50.02 Parts 2, 3, 4, 5, and 6. These fieldbus standards are also headed for final international approval.

Local Area Networks

Multi-vendor equipment always exists in large computer control systems. This equipment is interconnected through LANs. Communicating between these various vendors' systems presents major problems. The solution is for communication to be based on a common protocol. Several vendor proprietary protocols exist, but today no single standard appears likely to emerge. With different LANs using different protocols, interconnecting devices such as repeaters and layered gateways are used to interconnect LANs.

15-3. Direct Digital Control (DDC) Systems

Soon after computers began to be used in process control systems, they began to be used as the "controller" in a number of control loops. Typically, multiplexers were used to time-share the computer among the loops. The controller in a conventional analog feedback system performs rather routine calculations on the error signal and determines what the value of the manipulated variable should be. The feedback controller is a special-purpose analog computer. As digital computers developed in capability and availability, it became apparent that they could be used to perform the calculations that were being done by analog feedback controllers. In effect, the idea evolved that a single digital computer could be time-shared among a number of different feedback control loops. This is illustrated in Fig. 15-1.

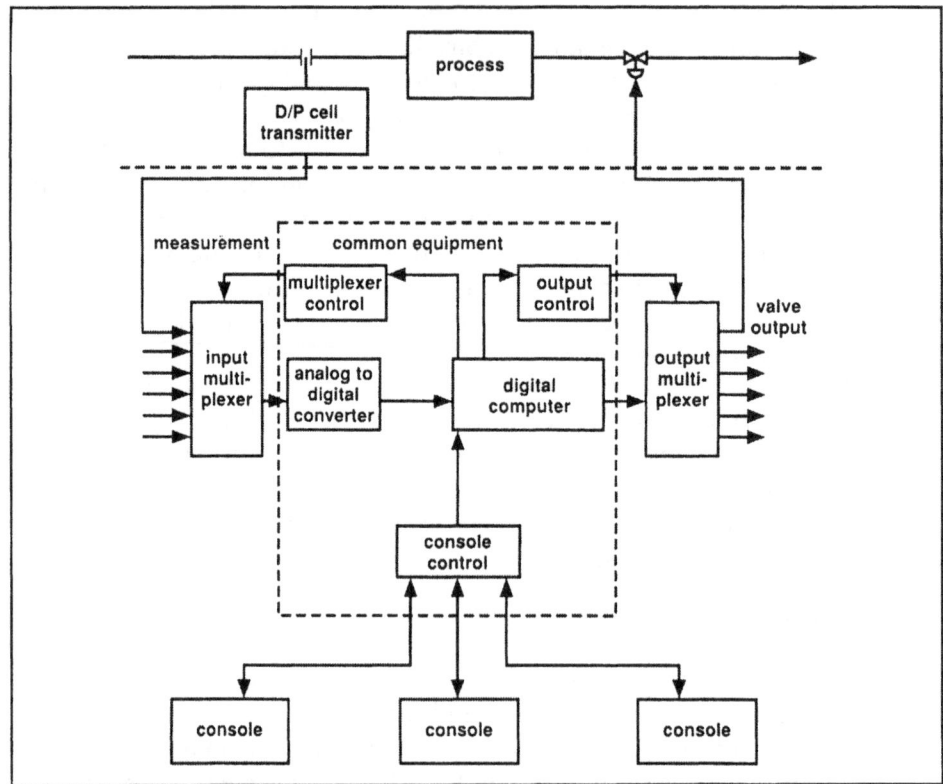

Figure 15-1. Direct Digital Control (DDC)

The computer has an input subsystem that make possible the collection of signals from the various sensors of the controller variables and other significant operating conditions of the plant. Using conventional feedback control equations or algorithms, the digital computer calculates the needed values of the manipulated variables in the plant, and the computer output becomes signals sent to the various final control elements. Basically, in this elementary concept of direct digital control, or DDC, the computer is used as a discrete equivalent to the analog hardware it has replaced. For the conventional three-mode PID controller, the computer equation used to calculate the manipulated variable is a discrete equivalent of the conventional three-mode algorithm. In finite difference form, it is as follows:

$$m_n = K_c e_n + \frac{K_c T}{T_i} \sum_{i=0}^{n} e_i + \frac{T_d K_c}{T}(e_n - e_{n-1}) + M_r \qquad (15\text{-}1)$$

where:

m_n = value of manipulated variable in the nth sampling instant.
e_n = value of the error at the nth sampling instant.
T = sampling time.
M_r = midrange adjustment

The other parameters are all as we defined them in Unit 9.

This particular form of the direct digital control algorithm is referred to as the position form of the control algorithm, and its output is the actual value of the manipulated variable, for example, a valve position. If this equation were rewritten for the m_{n-1} sampling instant and subtracted from m_n as given in the preceding equation, the result is as follows:

$$\Delta m_n = K_c(e_n - e_{n-1}) + \frac{K_c T}{T_i}e_n + \frac{T_d K_c}{T}(e_n - 2e_{n-1} + e_{n-2}) \qquad (15\text{-}2)$$

where:

$$\Delta m_n = (m_n - m_{n-1})$$

This is the change that is needed in the manipulated variable and is referred to as the velocity algorithm. The significant difference between the position algorithm and the velocity algorithm is that the velocity algorithm does not contain the term M_r. Therefore, the computer will provide a smooth transition from manual to automatic control; that is, we will have bumpless transfer.

The first ideas for DDC systems were based on the concept that one computer could be the replacement for many analog controllers, and the savings in analog hardware could therefore be used to purchase the digital computer. There are several problems with this as a sufficient economic justification, however. First, DDC systems tend to be more expensive, and programming costs quite often turn out to be large. Second, DDC systems require some analog backup hardware since operating personnel must be able to exercise effective control over the plant in the event of computer failure. In some cases, this analog backup becomes a complete analog control system. Reduction in manpower is a potential justification for DDC systems, but this rarely happens. Most processing units are already automated and operating with a minimum staff, and therefore few people can be eliminated. As a matter of fact, the presence of the computer frequently entails the presence of higher-caliber personnel, and personnel costs often increase rather than decrease.

The sum of extensive experience in DDC leads to the conclusion that if the computer is used solely to replace earlier levels of automation it is rarely justified. To say differently, the computer needs to be used to automate functions and operations that were not being automatically accomplished before by simple analog controllers. The truly remarkable capabilities of the computer should be used to solve operating problems and improve the optimization of the unit. If this can be done, that is, if the capability of the computer can be exploited, then DDC quite often becomes economically attractive. Many of the concepts outlines in the previous two units, for example, feedforward control, dead time control, multivariable control, and so on, can be conveniently implemented using DDC. To accomplish many of these with conventional analog hardware would be difficult or impossible. The techniques we will discuss in Unit 16, such as statistical process control and artificial intelligence, make the need for digital capabilities even stronger. One DDC approach is to implement direct digital control only for those loops that yield significant and obvious improvement in control performance. The implication of this is that conventional analog control systems should be left on the loops where conventional feedback control is envisioned. This is a type of "hybrid" approach to process control because it involves mixing digital capabilities with conventional analog capabilities.

As digital systems get more capable, less expensive, more secure (redundant), and as users become more familiar with them, they are replacing more and more of what was once an analog world.

15-4. Supervisory Control Systems

When you observe the conventional fluid process unit, you see that most plants operate with large numbers of feedback controllers, and occasionally you encounter a particular loop with a specific bit of feedforward control or other sophisticated or advanced control installed. But even with so extensive a control system in place, closer inspection reveals that the supervisory function of the plant is usually only modestly automated, if at all. As an example, the individual plant operators or plant supervisory personnel make the determination of all of the set points for the feedback and feedforward controllers. They do this based on past experience and some elementary calculations. Experience has shown that quite often operating personnel undertake this responsibility in a very crude, very conservative, and very slow fashion. The result is that most plants are not operated at their optimal level.

The basic objective of a process operation is to produce financial return (profit) on investment. This economic return depends upon a number of factors, and the operating strategy is not always clearly perceived or broadly understood by all of the decision-making people involved. A plant is a complex, highly interactive entity, and many times the optimum operating strategy can be determined only through sophisticated calculations. Clearly, the digital computer should be able to help perform such calculations in a real-time, optimal way. This concept is illustrated in Fig. 15-2.

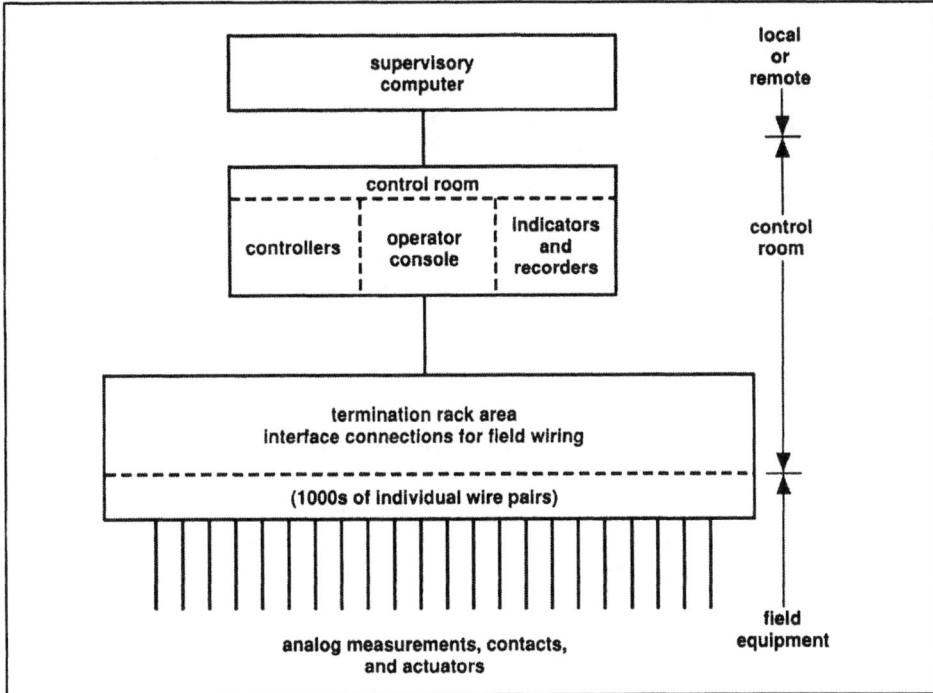

Figure 15-2. Supervisory Control System

The computer is able to measure significant amounts of information about actual operating conditions in the plant. Additional information may be provided to the computer, including the following:

- the cost of raw materials and utilities
- the value of products
- the composition of various raw materials, products, and intermediate streams
- the current values of variables within the process
- constraints on the operation of the process
- specifications on products

A model of the process can be used to relate these various factors, including providing a complete picture of plant economics. This model can then be optimized to determine the best operating strategy for the unit. The result is a specific indication of the optimum or most desirable values for the set points of all the individual controllers within the plant. These set points may be provided simply as information to the operator; that is, the supervisory control system may be open loop. In other cases, the set points for the controllers may be set directly via a hard-wired system. This is referred to as *closed-loop supervisory control*.

The installation of supervisory control systems has proved to be a very attractive economic concept in process control. It illustrates the fact that the supervisory control function was not automated before digital computers entered the scene, and it often was being done less efficiently than is possible with computers. The economic justification for supervisory control systems is based upon the prospect that the control system will produce improvements in process operations. There are, therefore, a number of situations in which supervisory digital control systems are prime candidates for installation. These include

- plants with large throughput
- very complex plants
- plants subject to frequent disturbances

The major difficulty in installing supervisory control systems is that mathematical models of plants are seldom available beforehand, and significant engineering and technical investigation are therefore required before the system may be installed and placed into operation. The incentive to do so is clearly present, however, because of the significant benefits that can accrue.

15-5. Control System Structure

Distributed architecture is currently a very popular structure for control systems. The name implies that actual process control and automated process management are functions that, in fact, are distributed throughout the plant. They are not concentrated in a specific geographic location called a "control room," nor are they inherently concentrated in a single piece of hardware. (The specific characteristics of process control and process management will be discussed in more detail in Unit 16.)

Since the need for control and management is distributed, it is logical to distribute the hardware and the capability to accomplish these two functions. Focusing on the structure question opens our thinking about the various control architecture structures that are possible. Fig. 15-3 shows five different structures for establishing a system to accomplish process control. Fig. 15-3(a) shows the hierarchical system that was the nature of most early digital control systems. Fig. 15-3(b) shows a general network. When compared to a hierarchical system or other similarly specialized organization, such a general network does allow for more redundant and reliable communications among the various components. The overhead cost of such systems, however, makes them less attractive for normal use.

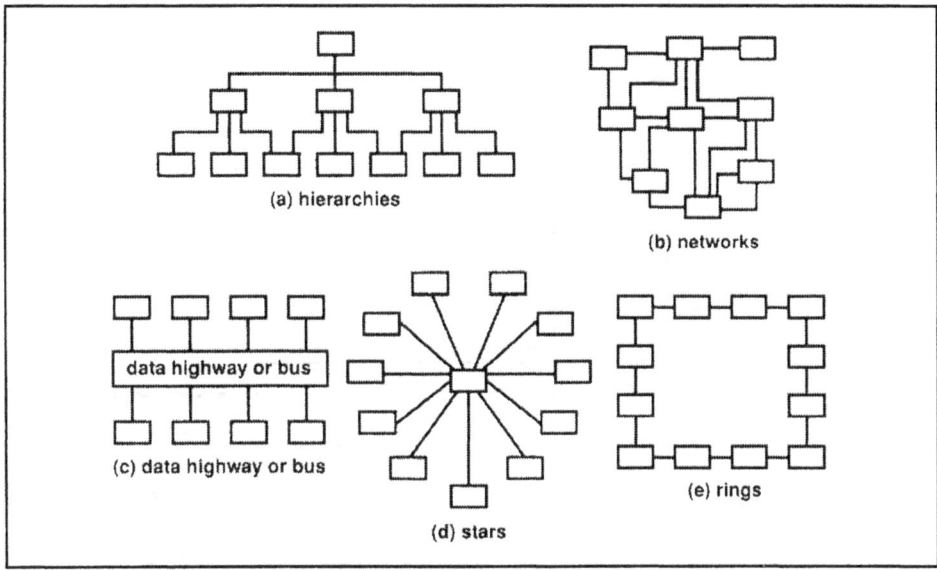

Figure 15-3. Possible Organizations for a Control System

Because of space limitations, we will make no special comment about star or ring systems, though the latter are used in some distributed control system installations. The data highway or bus system shown in Fig. 15-3(c) is the most popular system for implementing distributed control. One can argue that the system is fairly highly centralized about the highway itself, but making the highway itself secure offsets this.

As just stated, for many years all process automation systems had a very simple monolithic form or configuration. Analog signals from individual measurement sensors were transmitted to a central control room, and control signals to the actuators were returned in analog form. This basic configuration persisted effectively for three decades—from the time of the first centralized control rooms until the late seventies. Most of the early systems were pneumatic in nature, and as electronic control systems came on the scene they tended to duplicate the architectural configuration of pneumatic systems.

The basic concept of this type of centralized control configuration is shown in Fig. 15-4. In such installations, hundreds—and even thousands—of control loops are directly tubed or wired into the control room. When digital control first came into use, much of the discussion was about direct digital control. The basic architectural configuration remained monolithic and totally centralized. All information processing and all computation, whether done by analog techniques or digital ones, were done within the control room or in stand-alone, field-mounted controllers.

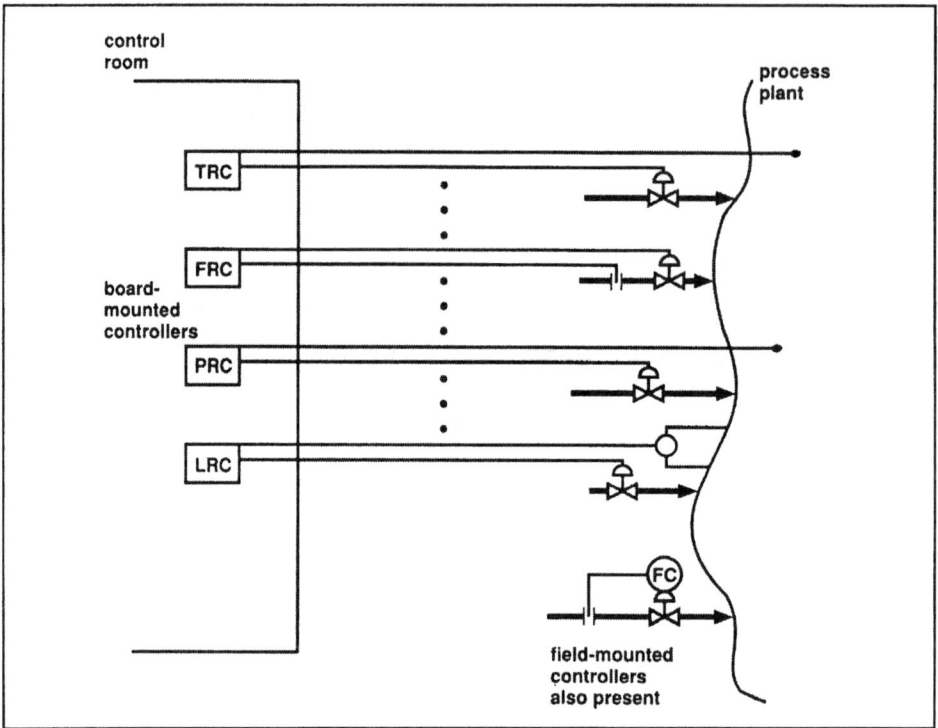

Figure 15-4. Centralized, Single-Loop Architecture

In one sense, this type of configuration is similar to the star-type configuration shown in Fig. 15-3(d). In a true star configuration all communication is through the central system, but in actual application there can be elaborate communications in and among the various loops.

15-6. Distributed Control Systems (DCS)

It is clear that architectural design has moved away from the simple, individual loop arrangements for process control and toward interconnected distributed systems with elaborate interconnecting communications. This communication is usually not broadband, but it does tend to be over Ethernet. The idea of a distributed system, which was first introduced in the mid-1970s, is illustrated in Fig. 15-5. In distributed control, the individual feedback controllers for each process loop are removed from the control room location and placed closer to the field sensors and/or actuators. When this is done, of course, there arises a need for a significant communication link, for example, a digital bus or data highway that connects the individual controllers with operators, computers, consoles, and displays. The control loops become physically shorter and therefore less vulnerable to noise or damage.

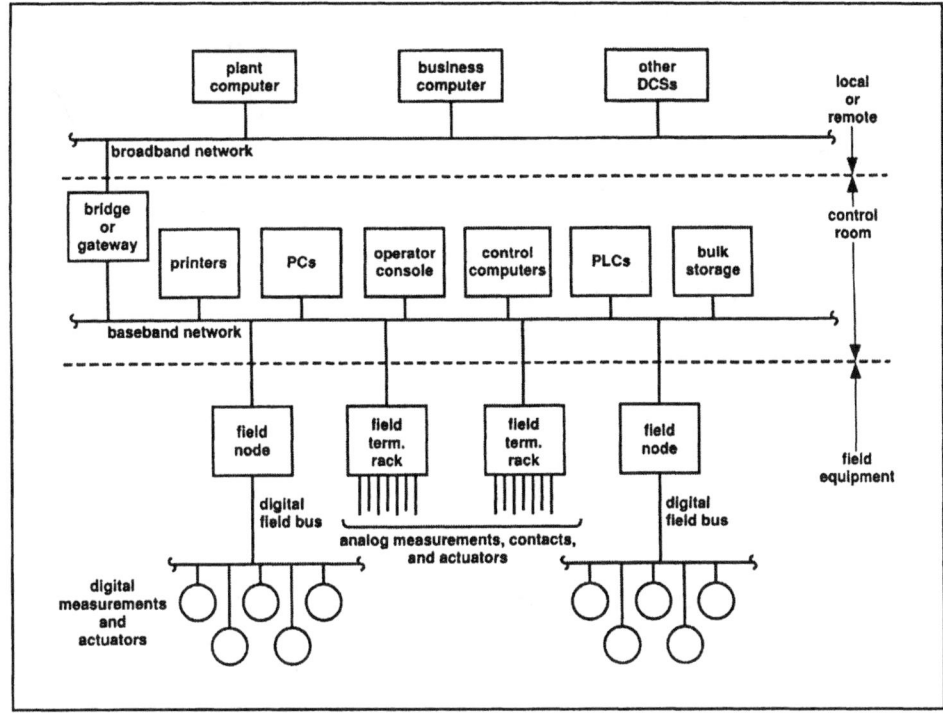

Figure 15-5. Distributed Control System (DCS)

Although the communication link might be lost, this basically represents a loss of operator intelligence because the individual field controllers continue to operate locally. Systems of this type can be implemented with either analog or digital controllers and even the pneumatic controllers. However, to date the mostly digital hardware approach prevails and will continue to do so in the future.

In the DCS approach, the operator in the central control room has access to all controller data such as set points, process variable measurements, controller output signal levels, and so forth. Sophisticated displays are also available, and the supervisory or management function—including all of the potential for advanced management information, optimization, and supervision—are easily implemented within the control room itself. Distributed control systems contain field nodes and termination racks, thus reducing the high cost of field wiring and allowing for more accurate signal transmission. Since so many sensors now contain microprocessors, digital signals can be transmitted without any use of 4–20 mA DC transmission signals. In 4–20 mA DC transmission systems, field devices are needed to do analog-to-digital conversion.

It is a statement of the obvious, but in a DCS system the fields truly become critical. Several vendors have proprietary fieldbuses, but it is clear that the standard being prepared by an international consortium at the time of this writing is needed. It is also obvious that the bus needs to be fast and needs to be secure. As we noted earlier, this is usually achieved through redundancy. In DCS, the software for operator input-output (I/O), the basic control algorithms, and the system operation software are embedded in the various elements of the system. Not only can the user call forth PID control, but cascade, ratio, feedforward, and other advanced algorithms are also readily available. Software for handling, storing, and displaying data is also provided, as is a high-level application language for establishing advanced control and process management functions. The structure of the software may be oriented toward either continuous or batch operations.

Although we have not really discussed it here except in passing, the process control system may be connected to business systems associated with the plant. This option entails some major economic justification issues.

15-7. Single-Loop or Multi-loop Structures

The principal thrust of the first ten units of this book was on a single loop (Ref. 8). In this unit we will shift the focus toward large systems and their architecture. But these large systems can be structured as simple aggregations of many single-loop controllers (SLCs) or as multi-loop controllers (MLCs), in which the loops are integrated more fully with one another as well as with the rest of the control system. Let's explore the differences between the two and the advantages and disadvantages of each.

The differences between SLC and MLC can be understood best by inspecting their historical roots. SLCs grew out of the pneumatic and then the electronic analog controllers. These were designed and installed as single loops. In time, the electronics tended to move the controllers from field-mounted locations to a more centralized control room where the environment was more hospitable to the electronic devices. As microprocessors came along, they tended to simply replace and duplicate the analog loops.

MLCs, on the other hand, had their beginning in the direct digital control type of computer control system. As microprocessors emerged, it became practical and desirable to use one as the front end of a computer control system. The data acquisition, as well as basic feedback control, could be done in the microprocessor. The advantages were better response time and, since the system was more compartmentalized, minimized impact when a component failed.

Table 15-1 lists the general characteristics of SLCs and MLCs. One comment of special note is that the differences between SLCs and MLCs tend to diminish as the capabilities of microprocessors and advances in electronics in general have made their impacts on process control systems. To put this differently, the two types are both moving toward one another, and the distinctions tend to blur.

The adoption of MLCs is without question the trend for large computer-based systems since, as you might expect, they represent more powerful

	Single-Loop Control	Multi-Loop Control
Cost per loop	Has lower cost per loop for small systems.	Costs more per loop for smaller systems. But larger systems can have lower installed per-loop costs.
Entry cost for first loop	Has lower entry cost if panel space and mimic boards are present.	Has higher entry cost.
Ability to network large systems	Limited.	Superior.
Person-machine interface functionality	Limited. New technology and PCs provide some solutions.	Superior.
Person-machine interface cost	Low. Is limited to a smaller number of points. Is closed to data that originates outside of system.	High. Has a higher capacity. Can access more plant-wide data, not just process control data.
Controller functionality	Limited. Powerful micros and more memory give some solutions.	Superior.
Expansion and upgrade potential	Is expandable one loop at a time. Has limited ability to expand to large systems and significant new functionality. Update technologies give better expansion and upgrade potential.	Allows expansion to proceed in large chunks at higher incremental cost, but a lower cost per loop.
Ability to integrate other devices	Is difficult to integrate external devices.	Is easy to integrate with higher-level or intelligent external devices.
Modularity and impact of a failure	Has the ability to limit the impact of most failures.	Failures in nonredundant multi-loop controllers could affect a significant number of loops.
Links for high-communication throughput	Remains as stand-alone device or resides on a simple serial link. An interface or bridge may provide higher-speed communications.	Resides on a redundant high-bandwidth communications link. This link might be used for controller-to-controller information and for controller-to-other devices information.

Table 15-1. The General Characteristics of Single-Loop and Multi-loop Controllers (Ref. 8)

controller capabilities, better networking characteristics, and better operator interface characteristics. Reliability is always a concern, but this is usually solved with redundancy. Table 15-2 shows the architectural characteristics of both SLCs and MLCs.

	Single-Loop Control	Multi-Loop Control
Interaction between loops	Acts as stand-alone entities.	Is highly interactive.
Operator interaction	Requires little or no processing beyond close-the-loop control.	More complex processes may require more operator interaction.
Permanency of loop configurations	Assumes permanent with little or no need to reconfigure or change.	Assumes that configuration may change significantly and frequently.
Interface to outside devices	Accomplishes interface via analog signal or status contact.	Accomplishes interface via digital communications with system-wide access.
Plant operation	Assumes that plant operation will be done at a higher-level device.	May assume that plant operation will be done at the multi-loop controller person-machine interface.
Treatment of process data	Treats data as property of the originating loop, not a system-wide resource.	Treats data as property of the entire system, including field devices, attached intelligent devices, person-machine interfaces, and higher-level devices.

Table 15-2. Architecture of Single-Loop and Multi-loop Controllers (Ref. 8)

15-8. Sequential or Batch Control

It should be noted that up to this point we have concentrated totally on continuous control and have ignored so-called batch control or sequential control. Since this entire field of process control is discrete and event ordered (think of Mom's apple pie being made by recipe in the kitchen!), it is not given to generalized theoretical treatment, hence, we have not focused on it thus far. Sequential or batch control is most important, however, and there are several ANSI/ISA standards on the subject. For example, ANSI/ISA-S88.01-1995 Batch Control Models and Terminology. The standards focus on protocols, however, and are procedure oriented. If the reader is interested in this kind of control, we refer him or her to these standards for detailed treatment.

It is also appropriate to note that programmable logic controllers (PLCs) are often used to implement sequential control systems. However, their hardware and software capabilities as well as those of distributed control systems are converging, so they are increasingly competitive with one another. This is especially true for larger systems.

REFERENCES

1. Kenna Amos, "1999 Outlook: From the Control Room to the Field," *InTech*, vol. 46, no. 1 (January 1999), pp. 38-45.

2. L. T. Amy, *Automation Systems for Control and Data Acquisition* (ISA, 1992).

3. Richard H. Caro, "SP50 Chair Explains How Buses Relate," *InTech*, vol. 43, no. 11 (November 1996), pp. 18-19 (last of a four-part series in 1996).

4. Albert A. Gunkler and John W. Bernard, *Computer Control Strategies for the Fluid Process Industries* (ISA, 1990).

5. Samuel M. Herb, *Understanding Distributed Processor Systems for Control* (ISA, 1999).

6. William R. Hodson, "How Fieldbus Will Affect DCS Architecture," *InTech*, vol. 43, no. 11 (November 1996), pp. 50-53.

7. Joseph LaFauci, "Choose the Right Control System," *Chemical Engineering Progress*, (March 1998), pp. 55-60.

8. Tom Wallace, "Single- and Multi-loop Vie for Control," *InTech*, vol. 37, no. 9 (September 1990), p. 180.

Unit 16:
New Directions
for Process Control

UNIT 16

New Directions for Process Control

The tools for process control—both the hardware/software and the knowledge—have been improving and expanding at a very rapid rate. Especially given the explosive changes brought about by digital computer capabilities, entire new directions of development and application are being created. This unit examines some of these phenomena.

Learning Objectives — When you have completed this unit, you should:

A. be able to describe the difference between process control and process management

B. be able to state the basic approach of statistical process control (SPC) and the tools used therein

C. be able to describe artificial intelligence/expert systems and what they can do

D. be able to discuss the concept of computer-integrated manufacturing (CIM)

16-1. Process Control and Process Management

This book has been devoted principally to specific control techniques and hardware, with additional attention given to the overall structure of process automation. In Unit 2, the distinction between process control and process management was briefly presented. It is now appropriate to expand on this subject further.

When a process is automated, the first general efforts are toward measuring process variables, and simple hardware automation techniques are used to establish basic control over the operation of the plant by controlling a few specific variables. As process control functions become more elaborate and higher levels of plant automation are undertaken, there begins a shift of focus toward automating more and more of the management of the plant. It is desirable to conceptually separate the process control function from the process management function. Although quite often the distinctions between these two basic areas of automation are blurred and diffuse, by appreciating some of the distinctions between them you can gain a better overall understanding of total plant automation.

Fig. 16-1 shows a general outline of the progression of plant automation control systems. The vertical axis in Fig. 16-1 shows increasing levels of control automation while the horizontal axis shows increasing levels of process management. It is clear that the major initial thrusts of automation are toward plant control, but subsequent and advanced control systems begin to focus more and more on automating the management function.

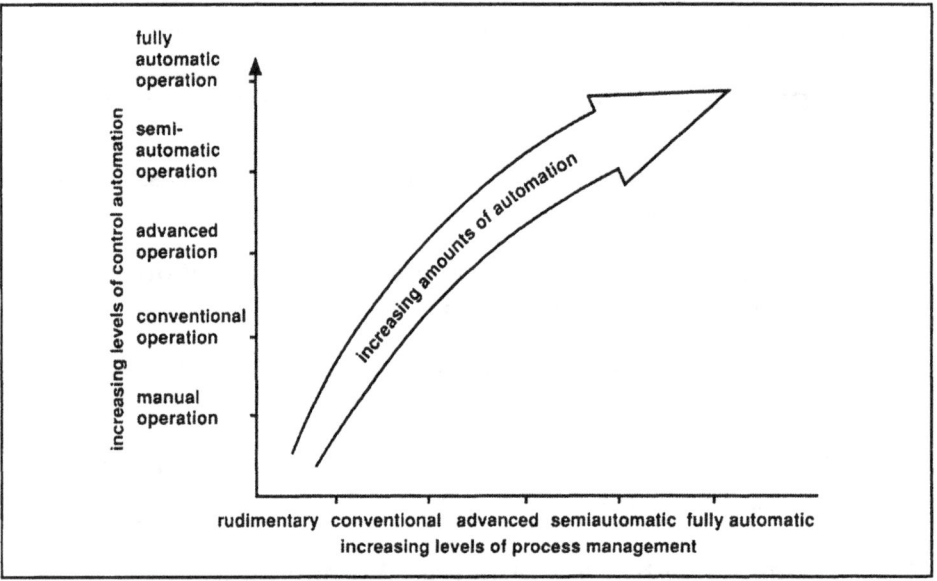

Figure 16-1. Progression of Advanced Control Systems (courtesy of the Foxboro Company)

16-2. Specific Characteristics of Process Control Automation

The vertical scale in Fig. 16-1 describes increasing levels of control automation. These five general levels of control automation have specific individual characteristics, which are characterized in Table 16-1. As the change is made from manual operation to fully automatic operation, the various characteristics of process control become increasingly automated and more sophisticated.

characteristics / levels of control automation	manipulation of process actuators			control law used during normal operation of plant	method of controller tuning	human monitoring and intervention	alarm condition analysis	measurements used and variables computed	plant scope
	under normal operation	during start-up, shutdown level chgs.	during emergency conditions						
manual operation	human	human	human	written decision rule	NA	NA	human	as necessary	independent unit
conventional operation	automatic	↓	key actuators automatic	on-off, 3-mode, cascade	human	automatic manual, emergency trip	↓	↓	↓
advanced operation				above feed-forward, noninteracting	some key loops tuned automatically		some alarm analysis displayed		integration of single units
semi-automatic operation		semi-automatic	all actuators automatic	above coordinated	all key loops automatic	↓	some automatic action		integration of multiple units
fully automatic operation	↓	fully automatic	↓		fully automatic	emergency trip	fully automatic	↓	plant treatd as single entity

Table 16-1(a). Characteristics of Increasing Levels of Control Automation (courtesy of The Foxboro Company)

As a process becomes more fully automated, more and more technology must be brought into use. In Table 16-2, the five increasing levels of control automation from Fig. 16-1 are presented in terms of the functional characteristics of the control technologies they require.

functional characteristics of control technologies required \ Increasing levels of control automation desired	manual operation	conventional operation	advanced operation	semi-auto. operation	fully auto. operation
measurements					
basic measurements (T, P, L, F)	X	X	X	X	X
advanced (composition, vibration)			X	X	X
complex (product quality)				X	X
two state detectors				X	X
control decisions					
on-off controller		X	X	X	X
3-mode controller		X	X	X	X
model-based controller			X	X	X
alarm diagnostic routine			X	X	X
automatic controller tuning routine			X	X	X
logic sequencing				X	X
display and reporting					
display and recording of variables	X	X	X	X	N/A for normal operation
annunciate alarms		X	X	X	
alarm diagnostic messages			X	X	
status displays				X	
control models					
simple heat & mat. balance models			X	X	X
dynamic process models				X	X
detailed safety models				X	X
variable manipulation					
actuators (basic)	X	X	X	X	X
actuators (increased accuracy)				X	X
actuators (on-off)				X	X
actuators (increased reliability)					X

Table 16-1(b). Function Characteristics of Control Technologies Required (courtesy of The Foxboro Company)

Tables 16-1 and 16-2 show the designer what is possible in process control automation and the functional characteristics of the associated technologies. It is clear that some industries are more sophisticated in the use of control automation than others, and, of course, within specific industries there are wide variations from plant to plant and from company to company. It is clear, however, that the strong trend is toward increasing levels of control automation and the increased application of the technologies associated with it.

16-3. Specific Characteristics of Process Management

We have made a distinction between process control and process management, and in Fig. 16-1 we showed the typical evolution of increased levels of total automation. In the last section we also presented some of the specific details of increased levels of process control. Let's undertake a similar presentation with respect to process management. Table 16-3 shows the five levels of process management (as defined in Fig. 16-1) and the characteristics associated with them.

Just as with process control, the five levels of process management have associated with them various technology requirements, which are illustrated in Table 16-4. The general trend and the analogy to process control are clear.

character-istics / levels of progress mgt. needed or desired	data gathering and retention			use of management and process information			
	data gathering and conversion to machine readable form	data type entered into process management system	data retention	display of performance variables	basis for decisions	degree of learning	plant scope
rudimentary process mgt.	manual gathering	none	logs, records	none	human intuition	human	independent single unit
conventional process mgt.	manual gathering & conversion	some process	some system retention	some necessary	human judgment		
advanced process mgt.	some automatic gathering & conversion	most process some business	most system	most necessary	some automatic decisions	some models updated routinely	some integration of single units
semiautomatic process mgt.	most automatic	all process most business		all necessary	most automatic	semi-automatic updating	integration of multiple units
fully automatic process mgt.	fully automatic	all data	all system	no variables displayed	fully automatic	fully automatic	plant treated as single entity

Table 16-2(a). Characteristics of Increasing Levels of Process Management
(courtesy of The Foxboro Company)

functional characteristics of process mgt. technologies required / Increasing levels of process management desired	rudi-mentary	conven-tional	advanced	semi-auto-matic	fully auto-matic
Information gathering and conversion					
process data gathering	X	X	X	X	X
process data entry		X	X	X	X
business data gathering and entry			X	X	X
decision analysis techniques					
material and energy balance		X	X	X	X
computation of performance variables		X	X	X	X
exception identification		X	X	X	X
inventory analysis			X	X	X
scheduling			X	X	X
optimization algorithms			X	X	X
forecast systems				X	X
process diagnostics				X	X
display and reporting					
data logging	X	X	X	X	N/A for normal operation
trend displays		X	X	X	
exception reporting		X	X	X	
receipts, orders, shipments display			X	X	
pattern displays			X	X	
user-directed hierarchical display			X	X	
decision models					
safety and environmental model		X	X	X	X
material balance model		X	X	X	X
energy balance model		X	X	X	X
inventory model (e.g., EOQ)			X	X	X
scheduling model			X	X	X
optimization model (e.g., LP, NLP)			X	X	X
distribution and transportation model			X	X	X
forecast model (e.g., econometric)				X	X
diagnostic model				X	X
process management directives					
operating guides		X	X	X	X
plant operating conditions			X	X	X
raw material selection			X	X	X
shipping & distribution instructions			X	X	X
utilities management			X	X	X
maintenance schedules			X	X	X
raw material ordering				X	X

Table 16-2(b). Functional Characteristics of Process Management Technologies Required (courtesy of The Foxboro Company)

Using this overall model for process automation, the individual practitioner can better envision how to automate his or her individual project.

16-4. Computer-Integrated Manufacturing (CIM)

The term *computer-integrated manufacturing* (CIM) was first used in discrete manufacturing processes to describe factory-floor computers used to control operations. The use of the term has gradually expanded until today it is used in all types of manufacturing operations, including the fluid process industries, to mean the complete integration of manufacturing computers, communications networks, and process control systems into all manufacturing functions. This is illustrated in Fig. 16-2.

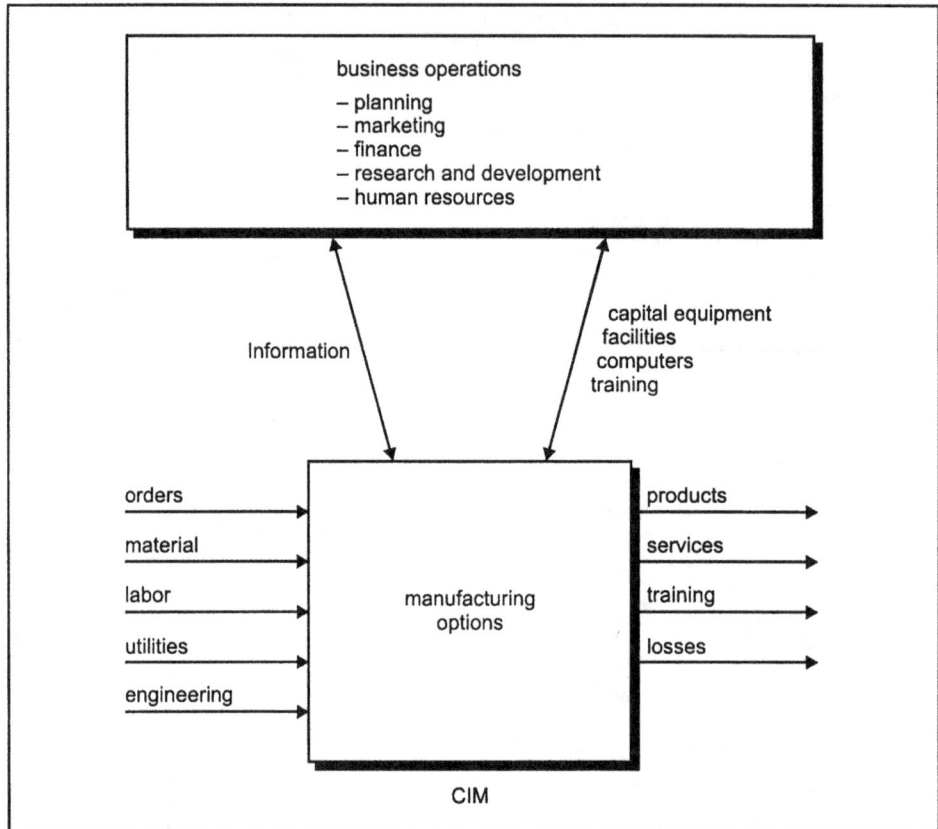

Figure 16-2. CIM Covers All Manufacturing Operations

One characteristic of the development of CIM is a seemingly endless array of buzzwords and acronyms to describe the various features of the new technology and its concepts. Some of the more common of these are given in Table 16-5.

CAD	— Computer-Aided Design
CAE	— Computer-Aided Engineering
CAM	— Computer-Aided Manufacturing
CIM	— Computer-Integrated Manufacturing
DDC	— Direct Digital Control
DNC	— Direct Numerical Control
FMS	— Flexible Manufacturing System
JIT	— Just In Time
LAN	— Local Area Network
MAP	— Manufacturing Automation Protocol
MRP	— Manufacturing Resources Planning
OSI	— Open System Interconnect
PC	— Programmable Controller or Personal Computer
PID	— Proportional-Integral-Derivative (Control)
PLC	— Programmable Logic Controller
TOP	— Technical and Office Protocol
WAN	— Wide Area Network

Table 16-3. Common Acronyms

Two technologies have enabled CIM to become a realistic concept. The first is a reliable, multi-vendor communications network that integrates all manufacturing operations, and the second is easy and timely access to the data that is required by the different operations.

Clearly, there is a rapidly declining cost for and increased capability of the basic components of CIM. These trends show no sign of slowing. They have made it possible in the past few years to implement practical applications of control theory, which have been known for some time, and they have also allowed for the development of new and improved process control techniques. In a similar way, these increased capabilities now allow technical and business functions to be included as an integral part of plant operations. That is what CIM is all about. It certainly will happen more and more rapidly as the years unfold!

16-5. Perspective on Statistical Process Control and Statistical Quality Control

During the late 1980s, statistical process control (SPC) burst onto the process control scene with a nearly religious intensity. Until that time the focus of quality concerns regarding production was primarily on the final product, and the statistical techniques dealing with control of the final product were called statistical quality control (SQC) methods. In the 1990s the use of SPC has only broadened and accelerated. SPC provides a methodology by which quality can be consistently measured, and thus controlled, in each step of a manufacturing process. Achieving good measurement of quality makes it possible to know and document the effects that process improvements and modifications are having.

The process control world has encountered some major difficulties in applying SPC, however. Fundamentally, SPC has been demonstrated and most effectively utilized in discrete manufacturing. Discrete manufacturing is built around linear assembly lines while process control is most often practiced in complex, interconnected, highly nonlinear flow systems that are most often continuous. Terms quite common in typical SPC applications, such as "zero defects" or "no scrap," do not fit directly into the traditional concepts of process control. This illustrates just how substantial the difference is between the use of SPC in continuous process control and in a conventional discrete manufacturing process.

Some of these differences need to be noted. In process control, the focus is on material and energy balances and their systematic adjustment. The process is said to be "in control" when sensors that track these material and energy balances are at or near their target values or set points. In SPC, a process is "in control" when sampled process measurements are normally distributed and fall within a window defined as a set number of standard deviations about a mean. In some environments, the term in control is used even if the mean is significantly different from the target or set point for that particular measurement.

Standard SPC also assumes that there are no mechanisms that create auto-correlation in the variables being controlled. Conventional control loops essentially ensure that auto-correlation will occur.

The basic focus of SPC is on the monitoring and study of the process in a systematic and statistically meaningful way. In the near term, this focus is on the identification and measurement of process problems that might otherwise go undetected. In the longer term, the focus is on using these measurements to chart improvements. The net of this is that, short term, the SPC practitioner tends to patiently accept variations that are statistically normal and to vigorously pursue those that exceed what is statistically expected.

Conventional process control assumes that all process relationships are deterministic in nature. SPC, on the other hand, assumes nature is inherently random and cannot be modeled solely by deterministic cause and effect. Only probabilities and distributions can describe some things. Conventional process control focuses on process dynamics, while the focus of SPC is static. Another contrast is that while process control is reactive, SPC is proactive.

16-6. The Tools of Statistical Process Control

The terminology of SPC is specialized (see Table 16-6); so are the tools. The tools of SPC are control charts and graphs, which are often constructed by the unit operator but are then reviewed by a quality control team before an action is taken. Note that the operator/engineer is "in the loop" in SPC.

Terminology

Alternative hypothesis (H₁). The hypothesis that the process mean or variability has shifted (the opposite of the null hypothesis, H_0).

Assignable cause. A process factor that causes abnormal or excessive variation in the process.

Attribute. A *qualitative* characteristic of a work piece. Attributes include the classifications conforming and nonconforming.

Binomial distribution. The binomial distribution states the probability of getting x occurrences given n number of trials and p probability of an occurrence. It requires *sampling with replacement* or sampling from a very large population without replacement.

Consumer's risk. The risk of a type II error, concluding that the process mean or variance is stable when it has actually changed (the risk of not finding trouble when there really is trouble; see *Producer's risk*).

Natural (or inherent) variation. The natural or inherent variation present in every manufacturing process.

Nonconformance. A work piece that does not meet customer specifications. A nonconforming work piece must either be scrapped or reworked.

Nonconformity. A defect.

Normal distribution. The bell curve probability distribution that is typical of most experiments and manufacturing processes.

Null hypothesis (H₀). The hypothesis that the true process mean (μ) equals the target process mean (or the same for process variability, σ, or for the fraction defective, etc.; in general, the hypothesis that everything is actually as it should be, or is as expected).

Outlier. An unusual observation. There is a high probability that the outlier did not come from the general population but could have resulted from an unusual occurrence or a measurement error.

Poisson distribution. Describes arrival frequencies. It can approximate the binomial distribution for large sample sizes and small probabilities of occurrence.

Producer's risk. The risk of a type I error, concluding that the process mean or variability has changed when it has not. (The risk of finding trouble where none exists, or crying wolf; see *Consumer's risk*).

Standard normal deviate (Z). The distance of a number from the mean of a normal population in standard deviations.

Variable. A *quantitative* or numerical characteristic.

Table 16-4. SPC Terminology

The most common quality control chart is the Shewhart chart. To construct such a chart, an operator periodically measures a process variable and records it as shown in Fig. 16-3. Note the upper and lower control limits and the expected mean, all of which are determined by historical data. If a measurement falls without the control limits or if successive data points-- for example, seven data points--fall on only one side of the mean, this is a signal for attention. In SPC terminology, the process is "out of control" until the cause of the abnormal event has been identified and corrected. Note that SPC control charts are not intended to take the place of conventional process control loops. Instead, they can be used to complement, enhance, and even optimize conventional process control.

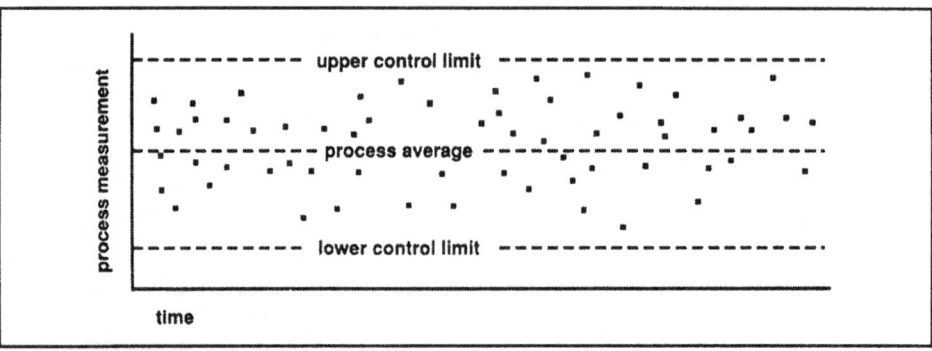

Figure 16-3. Shewhart Quality Control Chart

Just as there can be conventional process control that is advanced beyond simple PID feedback control, so SPC also has a range of more advanced tools.

When one is attempting to apply SPC to a process, an important early step is to identify the most serious problem to address. Pareto analysis is a tool that is helpful for separating, in a quantitative manner, the many trivial problems from the few critical ones. Fig. 16-4 shows a typical Pareto analysis. In this case study, it is clear that viscosity is the product characteristic that needs special attention. Once this determination has been made, the SPC team needs to list all possible reasons for the problem, and then these possible reasons are entered onto an Ishikawa or "fishbone" diagram. One is illustrated in Fig. 16-5. Note especially the headings for the various possible causes: "materials," "methods," "machines," "people," and "environment." These possible causes can then be prioritized to observe actual cause-and-effect relationships. As cause-and-effect data are collected, they can be entered onto SPC charts in order to document relationships and to communicate the problem. Over several months, you will gain a much more detailed understanding of the process and its operation, which should lead to a much higher quality level.

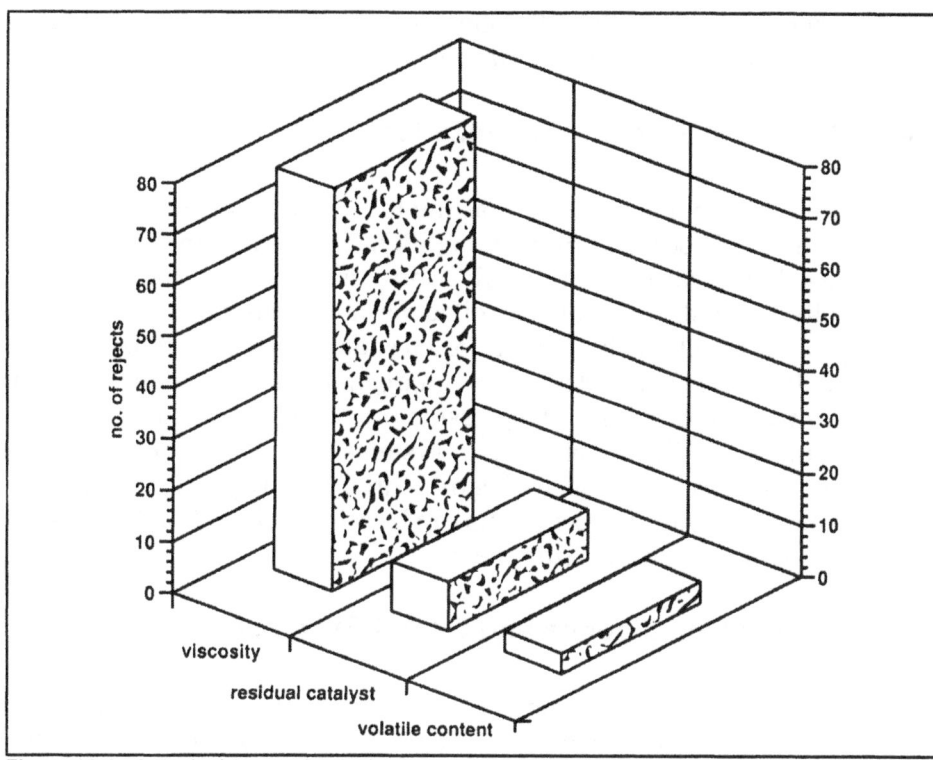

Figure 16-4. A Typical Pareto Analysis

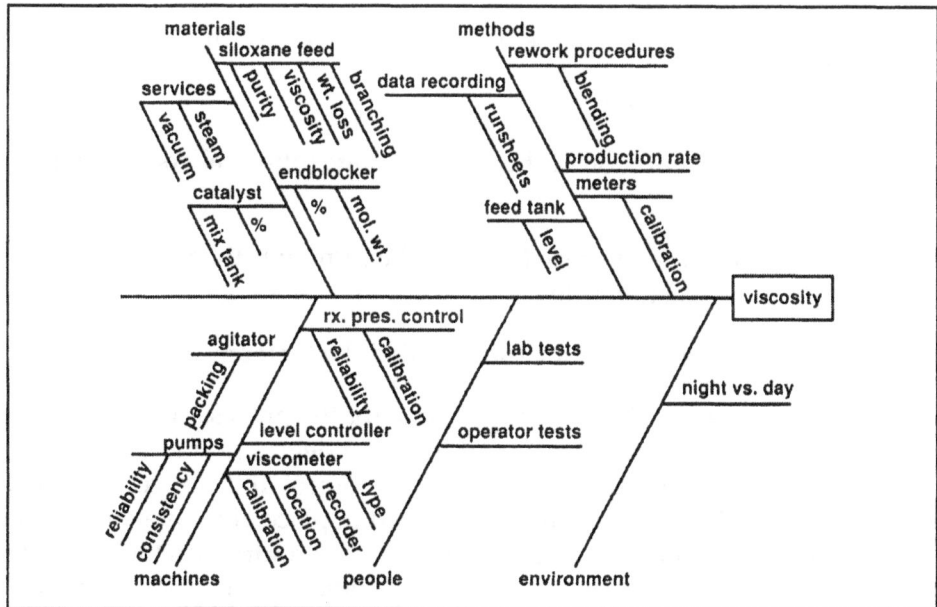

Figure 16-5. A Typical Fishbone Diagram

It should be noted that SPC tends to give discipline and procedure to what good engineers and good management should have been doing all along!

16-7. Statistical Process Optimization

A method that is currently starting to receive more and more attention is Statistical Process Optimization (SPO). This term implies a knowledge-based, self-learning, on-line SPC and process optimization method. Data is collected in the form of controlled variables, other process variables, and process results such as laboratory test data and/or cost data. These data are then used to build a process model that is statistical in nature, such as a multivariable, nonlinear regression model. The model can contain artificial intelligence and expert systems (see the next section) to improve (that is, help the model to "learn") through new process data. Then new values for controller set points can be selected based on the consequences that will occur from using them.

16-8. Artificial Intelligence and Expert Systems

Expert systems are a branch of the study of artificial intelligence (AI). Expert systems are computer programs that solve or help to solve problems in a specialized field at a level comparable to that of an expert in that field. Expert systems represent a very promising technology with some obvious potential for process control.

An expert system is made up of the following components:

- user interface

- representation of the problem state (often called the "global database")

- a knowledge base (often in the form of if-then rules—often called "production rules")

- the control regime (often called the "inference engine")

Expert systems show great potential for the process control and process management world because of the great degree of uncertainty and variability that is involved in process control. It is not yet clear if expert systems should, in the future, be used in place of or as a supplement to traditional equipment in real-time process control. As the systems develop, it is clear that the early applications will be at or near the man-machine interface and many will involve control loops that include a "man in the loop." Most expert systems in use today are fledgling systems and tend to be used off line for areas such as equipment condition, monitoring, and diagnosis, process advising and scheduling, alarm handling, hardware system diagnosis, control system turning, and control system configuration. While it is true that expert system-based process control has already been implemented, it is certainly also true that such implementations are of extremely limited scope and, to date, have been done on an experimental basis.

So far, expert systems have not fulfilled the expectations many people held for them. An obvious problem is that these systems are technologically complex, and the professionals who understand the technology for creating expert systems are quite often not those who understand the fields where their application is needed. Expert systems really are still in the development stage, and those functioning today are principally prototypes. Oftentimes, they have involved the use of significantly large resources and have been useful only to demonstrate the potential of the technology. Very few have penetrated real-world process control activities. To help solve this problem, a number of "development tools" have appeared in the marketplace to assist in the development of expert systems. These tools are designed to help create expert systems much more rapidly than would be possible if the programming itself were done in LISP, that is, the standard AI language used for the past two decades. These tools are also conceptual aids in that they help expert systems designers think about a problem in a way that fits the planned application. At this point, it is clear that the more generalized the expert system, the poorer the expert system. To say this conversely, the more precisely and narrowly defined the system, the more powerful it is when used in applications.

The cost of creating expert systems is dropping significantly because of the development tools just mentioned. These tools are often called "shells" because they contain the skeletal structure of an expert system without including the rules or knowledge that is specific to a particular application. This means that it is not necessary to have an artificial intelligence specialist in order to create an expert system. A good programmer or computer scientist can learn to create the systems. As a result, it is clear that expert systems will become a growing part of the supervisory control and process management world.

16-9. Neural Networks

The simplest definition of a neural network, which is more properly referred to as an "artificial" neural network (ANN), is a computing system made up of a number of simple, highly interconnected processing elements that process information by their dynamic-state response to external inputs (Ref. 11). They are loosely modeled after the neuronal structure of the mammalian brain. They are useful in process control applications to model, predict, control, and optimize process behavior.

Just as the human brain "learns" to recognize patterns, forecast outcomes, and respond appropriately, a neural net can be "trained" on process data to recognize pattern relationships between input and response observations. Neural net software programs are typically organized in layers, as shown in Fig. 16-6. These layers are made up of a number of interconnected "nodes" that contain an "activation function." Patterns of input data are presented to the network via the input layer, which in turn communicates to one or more "hidden layers" where the actual processing is done through a system of weighted "connections." The hidden layers then link to an "output layer" where the answer is an output response.

The hidden and output layer nodes process their incoming "signals" by applying factors ("weights") to them and then combining them. The resulting signal is then processed through a transfer function that controls the strength of the node's output. This is shown in Fig. 16-7.

Network processing continues through each layer until the network's combined response is obtained in the output layer. The number of nodes in a hidden layer, as well as the number of hidden layers in a model, are important characteristics of a neural network model. Transfer functions

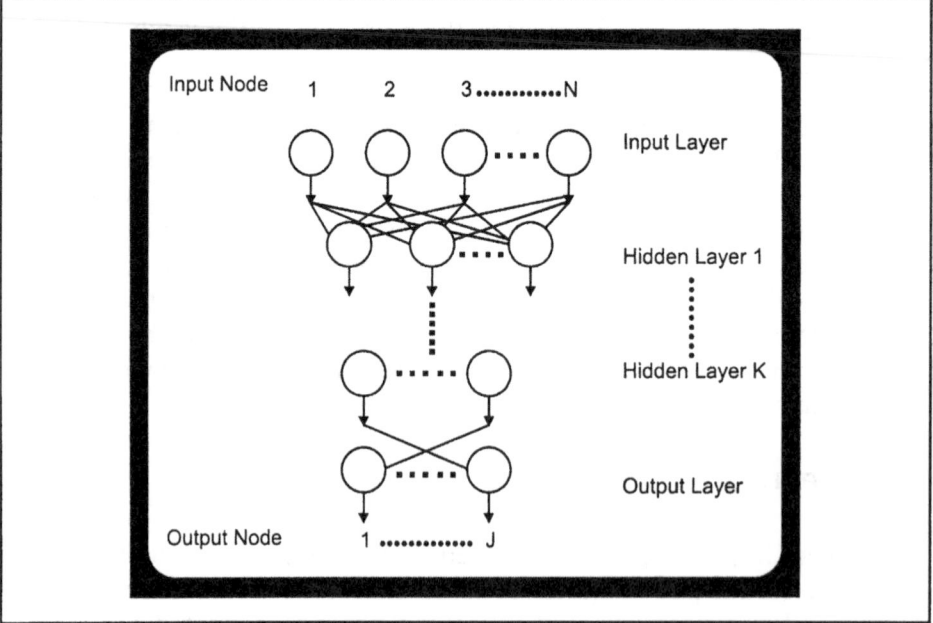

Figure 16-6. The Layers of a Neural Network

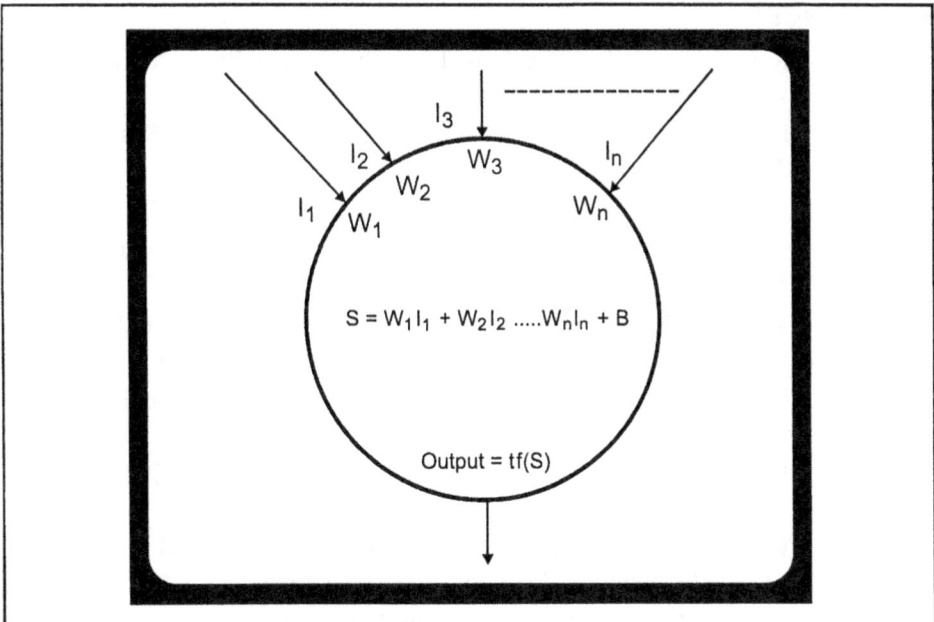

Figure 16-7. A Node in a Neural Network

serve to normalize each node's output signal, and they also offer each node the opportunity to affect the model in either a linear or nonlinear fashion, as dictated by the training algorithm.

All of this "training" of the neural net is done automatically on existing process data. Thus, the net "learns" a model of the process data. During this training, the network's response at the output layer is compared to a supplied set of training targets, the root mean squared error is measured and "back-propagated" through the network to improve the network's accuracy in predicting the training targets. Nodal weight factors are adjusted to drive the network's response error down to a minimum.

Neural nets are extremely powerful mathematical modeling tools, and they are thus useful to predict process dynamic behavior and to serve as virtual sensors. These models can also be inverted and used to control processes. These inverted models then can be optimized, and the result is dynamic, nonlinear, optimized control of most complex processes.

Neural nets are most powerful tools!

REFERENCES

1. Kenna Amos, "1999 Outlook: From the Control Room to the Field," *InTech*, vol. 46, no. 1 (January 1999), pp. 38-45.

2. P. S. Bauer, "Development Tools Aid in Fielding Expert Systems," *InTech*, vol. 34, no. 4 (April 1987).

3. John W. Bernard, *CIM in the Process Industries* (ISA, 1989).

4. Steven R. Block, "Improve Quality with Statistical Process Control," *Chemical Engineering Progress* (November 1990).

5. Martin Dybeck, "AI and SPC Team Up in Process Control," *InTech*, vol. 36, no. 12 (December 1989).

6. William R. Hodson, "How Fieldbus Will Affect DCS Architecture," *InTech*, vol. 43, no. 11 (November 1996), pp. 50-53.

7. Dick Johnson, "Ethernet Edges toward Process Control," *Control Engineering* (December 1998), pp. 66-72.

8. William Levinson, "Understand the Basics of Statistical Process Control," *Chemical Engineering Progress* (November 1990).

9. Y. Z. Lu, *Industrial Intelligent Control* (ISA, 1995).

10. Charles L. Mamzic, *Statistical Process Control*, Practical Guide Series (ISA, 1995).

11. Greg Martin, "Neural Network Applications for Prediction, Control, and Optimization," *ISA/1995 Advances in Instrumentation and Control*, vol. 50, pt. 23, p. 433.

12. N. S. Rajaram, "Expert Systems Development: Current Problems, Future Needs," *InTech*, vol. 34, no. 4 (April 1987), pp. 25-26.

Appendix A:
Graphic Symbols
for Process
Measurement
and Control

APPENDIX A

The Graphic Symbols for Process Measurement and Control

The basis for the graphical presentation of process control and measurement information is the following ANSI/ISA standard: S5.1-1984 (R 1992), "Instrumentation Symbols and Identification." The material presented in this appendix consists of excerpts from that standard and is used with permission.

Table A-1 provides the instrument line symbols given in the standard.

The standard also provides a system of alphanumeric codes or "tag numbers" to identify each instrument. The basic concept is for each instrument to be identified by a symbol, which is accompanied by a tag number. Fig. A-1 provides an example of a tag number.

	TYPICAL TAG NUMBER	
TIC 103		- Instrument Identification or Tag Number
T 103		- Loop Identification
103		- Loop Number
TIC		- Functional Identification
T		- First-Letter
IC		- Succeeding-Letters
	EXPANDED TAG NUMBER	
10-PAH-5A		- Tag Number
10		- Optional Prefix
A		- Optional Suffix
NOTE: Hyphens are optional as separators		

Figure A-1. Tag Numbers

Table A-2 provides the instrument identification letters that are used to construct the functional identification of tag numbers. The system that is used involves having the first letter designate the measured or indicated variable and using one or more succeeding letters to denote the functions performed.

ALL LINES TO BE FINE IN RELATION TO PROCESS PIPING LINES.

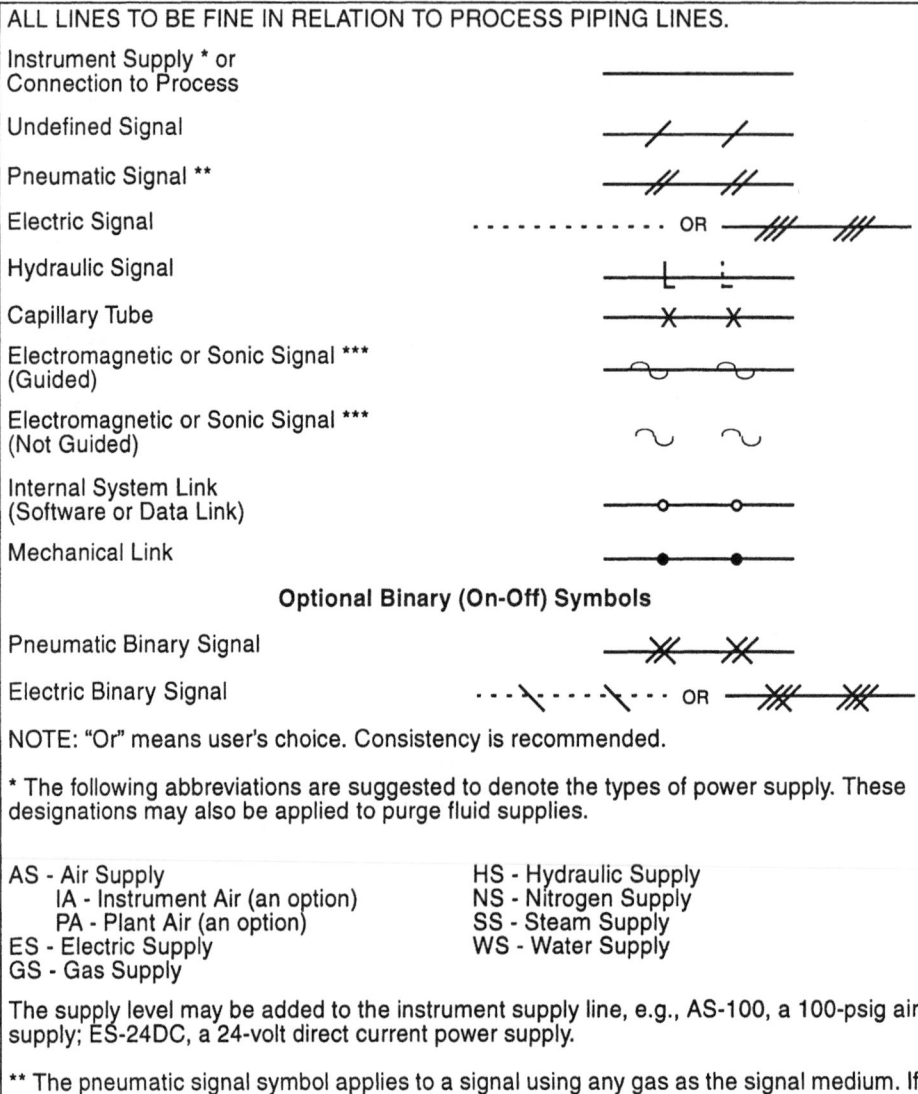

Instrument Supply * or
Connection to Process

Undefined Signal

Pneumatic Signal **

Electric Signal

Hydraulic Signal

Capillary Tube

Electromagnetic or Sonic Signal ***
(Guided)

Electromagnetic or Sonic Signal ***
(Not Guided)

Internal System Link
(Software or Data Link)

Mechanical Link

Optional Binary (On-Off) Symbols

Pneumatic Binary Signal

Electric Binary Signal

NOTE: "Or" means user's choice. Consistency is recommended.

* The following abbreviations are suggested to denote the types of power supply. These designations may also be applied to purge fluid supplies.

AS - Air Supply
    IA - Instrument Air (an option)
    PA - Plant Air (an option)
ES - Electric Supply
GS - Gas Supply

HS - Hydraulic Supply
NS - Nitrogen Supply
SS - Steam Supply
WS - Water Supply

The supply level may be added to the instrument supply line, e.g., AS-100, a 100-psig air supply; ES-24DC, a 24-volt direct current power supply.

** The pneumatic signal symbol applies to a signal using any gas as the signal medium. If a gas other than air is used, the gas may be identified by a note on the signal symbol or other wise.

*** Electromagnetic phenomena include heat, radio waves, nuclear radiation, and light.

Table A-1. Instrument Line Symbols

	FIRST-LETTER (4)		SUCCEEDING-LETTERS (3)		
	MEASURED OR INITIATING VARIABLE	MODIFIER	READOUT OR PASSIVE FUNCTION	OUTPUT FUNCTION	MODIFIER
A	Analysis (5,19)		Alarm		
B	Burner, Combustion		User's Choice (1)	User's Choice (1)	User's Choice (1)
C	User's Choice (1)			Control (13)	
D	User's Choice (1)	Differential (4)			
E	Voltage		Sensor (Primary Element)		
F	Flow Rate	Ratio (Fraction) (4)			
G	User's Choice (1)		Glass, Viewing Device (9)		
H	Hand				High (7, 15, 16)
I	Current (Electrical)		Indicate (10)		
J	Power	Scan (7)			
K	Time, Time Schedule	Time Rate of Change (4, 21)		Control Station (22)	
L	Level		Light (11)		Low (7, 15, 16)
M	User's Choice (1)	Momentary (4)			Middle, Intermediate (7,15)
N	User's Choice (1)		User's Choice (1)	User's Choice (1)	User's Choice (1)
O	User's Choice (1)		Orifice, Restriction		
P	Pressure, Vacuum		Point (Test) Connection		
Q	Quantity	Integrate, Totalize (4)			
R	Radiation		Record (17)		
S	Speed, Frequency	Safety (8)		Switch (13)	
T	Temperature			Transmit (18)	
U	Multivariable (6)		Multifunction (12)	Multifunction (12)	Multifunction (12)
V	Vibration, Mechanical Analysis (19)			Valve, Damper, Louver (13)	
W	Weight, Force		Well		
X	Unclassified (2)	X Axis	Unclassified (2)	Unclassified (2)	Unclassified (2)
Y	Event, State or Presence (20)	Y Axis		Relay, Compute, Convert (13, 14, 18)	
Z	Position, Dimension	Z Axis		Driver, Actuator, Unclassified Final Control Element	

Table A-2. Identification Letters

NOTE: Numbers in parentheses refer to specific explanatory notes on the following pages.

Notes for Table A-2

1. A "user's choice" letter is intended to cover unlisted meanings that will be used repetitively in a particular project. If used, the letter may have one meaning as a first-letter and another meaning as a succeeding-letter. The meanings need to be defined only once in a legend, or other place, for that project. For example, the letter N may be defined as "modulus of elasticity" as a first-letter and "oscilloscope" as a succeeding-letter.

2. The unclassified letter X is intended to cover unlisted meanings that will be used only once or used to a limited extent. If used, the letter may have any number of meanings as a first-letter and any number of meanings as a succeeding-letter. Except for its use with distinctive symbols, it is expected that the meanings will be defined outside a tagging bubble on a flow diagram. For example, XR-2 may be a stress recorder and XX-4 may be a stress oscilloscope.

3. The grammatical form of the succeeding-letter meanings may be modified as required. For example, "indicate" may be applied as "indicator" or "indicating," "transmit" as "transmitter" or "transmitting," etc.

4. Any first-letter, if used in combination with modifying letters D (differential), F (ratio), M (momentary), K (time rate of change), Q (integrate or totalize), or any combination of these is intended to represent a new and separate measured variable, and the combination is treated as a first-letter entity. Thus, instruments TDI and TI indicate two different variables, namely, differential-temperature and temperature. Modifying letters are used when applicable.

5. First-letter A (analysis) covers all analyses not described by a "user's choice" letter. It is expected that the type of analysis will be defined outside a tagging bubble.

6. Use of first-letter U for "multivariable" in lieu of a combination of first-letters is optional. It is recommended that nonspecific variable designators such as U be used sparingly.

7. The use of modifying terms "high," "low," "middle" or "intermediate," and "scan" is optional.

8. The term "safety" applies to emergency protective primary elements and emergency protective final control elements only. Thus, a self-actuated valve that prevents operation of a fluid system at a higher-than-desired pressure by bleeding fluid from

the system is a back-pressure-type PCV, even if the valve is not intended to be used normally. However, this valve is designated as a PSV if it is intended to protect against emergency conditions, i.e., conditions that are hazardous to personnel and/or equipment and that are not expected to arise normally.

The designation PSV applies to all valves intended to protect against emergency pressure conditions regardless of whether the valve construction and mode of operation place them in the category of the safety valve, relief valve, or safety relief valve. A rupture disc is designated PSE.

9.   The passive function G applies to instruments or devices that provide an uncalibrated view, such as sight glasses and television monitors.

10.   "Indicate" normally applies to the readout—analog or digital—of an actual measurement. In the case of a manual loader, it may be used for the dial or setting indication, i.e., for the value of the initiating variable.

11.   A pilot light that is part of an instrument loop should be designated by a first-letter followed by the succeeding-letter L. For example, a pilot light that indicates an expired time period should be tagged KQL. If it is desired to tag a pilot light that is not part of an instrument loop, the light is designated in the same way. For example, a running light for an electric motor may be tagged EL, assuming voltage to be the appropriate measured variable, or YL, assuming the operating status is being monitored. The unclassified variable X should be used only for applications which are limited in extent. The designation XL should not be used for motor running lights, as these are commonly numerous. It is permissible to use the user's choice letters M, N or O for a motor running light when the meaning is previously defined. If M is used, it must be clear that the letter does not stand for the word "motor," but for a monitored state.

12.   Use of a succeeding-letter U for "multifunction" instead of a combination of other functional letters is optional. This nonspecific function designator should be used sparingly.

13.   A device that connects, disconnects, or transfers one or more circuits may be either a switch, a relay, an ON-OFF controller, or a control valve, depending on the application. If the device manipulates a fluid process stream and is not a hand-actuated ON-OFF block valve, it is designated as a control valve. It is incorrect to use the succeeding-letters CV for anything other than

a self-actuated control valve. For all applications other than fluid process streams, the device is designated as follows:

- A switch, if it is actuated by hand.

- A switch or an ON-OFF controller, if it is automatic and is the first such device in a loop.

The term "switch" is generally used if the device is used for alarm, pilot light, selection, interlock, or safety.

- The term "controller" is generally used if the device is used for normal operating control.

- A relay, if it is automatic and is not the first such device in a loop, i.e., it is actuated by a switch or an ON-OFF controller.

14. It is expected that the functions associated with the use of succeeding-letter Y will be defined outside a bubble on a diagram when further definition is considered necessary. This definition need not be made when the function is self-evident, as for a solenoid valve in a fluid signal line.

15. The modifying terms "high," and "low," and "middle" or "intermediate" correspond to values of the measured variable, not to values of the signal, unless otherwise noted. For example, a high-level alarm derived from a reverse-acting level transmitter signal should be an LAH, even though the alarm is actuated when the signal falls to a low value. The terms may be used in combinations as appropriate.

16. The terms "high" and "low," when applied to positions of valves and other open-close devices, are defined as follows: "high" denotes that the valve is in or approaching the fully open position, and "low" denotes that it is in or approaching the fully closed position.

17. The word "record" applies to any form of permanent storage of information that permits retrieval by any means.

18. For use of the term "transmitter" versus "converter," see the definitions in Section 3.

19. First-letter V, "vibration or mechanical analysis," is intended to perform the duties in machinery monitoring that the letter A performs in more general analyses. Except for vibration, it is expected that the variable of interest will be defined outside the tagging bubble.

20. First-letter Y is intended for use when control or monitoring responses are event-driven as opposed to time- or time schedule-driven. The letter Y, in this position, can also signify presence or state.

21. Modifying-letter K, in combination with a first-letter such as L, T, or W, signifies a time rate of change of the measured or initiating variable. The variable WKIC, for instance, may represent a rate-of-weight-loss controller.

22. Succeeding-letter K is a user's option for designating a control station, while the succeeding-letter C is used for describing automatic or manual controllers. (See Section 3, Definitions.)

Table A-3 gives typical letter combinations used to denote the functions performed.

Fig. A-2 gives the general instrument or function symbols stipulated by the standard. For other symbols, such as for control valves, dampers, actuators and primary element symbols, refer to the ANSI/ISA standard.

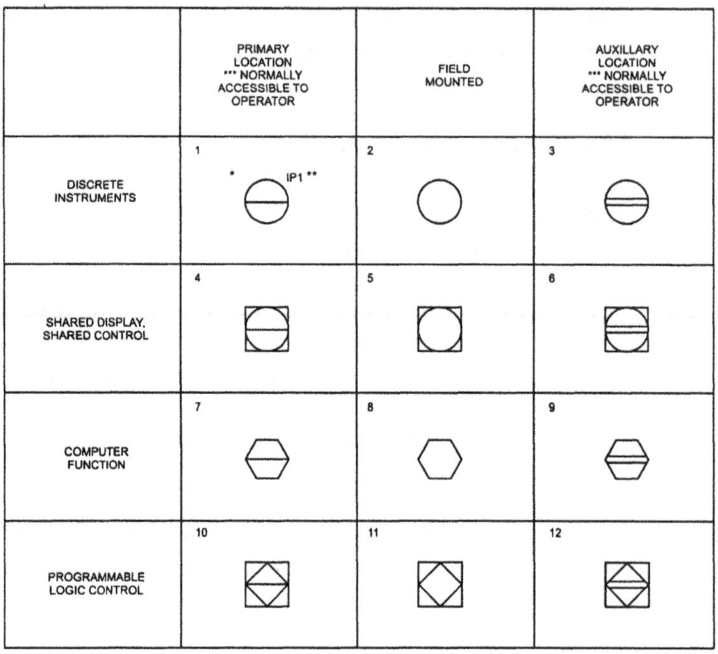

	PRIMARY LOCATION *** NORMALLY ACCESSIBLE TO OPERATOR	FIELD MOUNTED	AUXILLARY LOCATION *** NORMALLY ACCESSIBLE TO OPERATOR
DISCRETE INSTRUMENTS	1	2	3
SHARED DISPLAY, SHARED CONTROL	4	5	6
COMPUTER FUNCTION	7	8	9
PROGRAMMABLE LOGIC CONTROL	10	11	12

* Symbol size may vary according to the user's need and the type of document. A suggested square and circle size for large diagrams is shown above. Consistency is recommended.

** Abbreviation of the user's choice such as IP1 (Instrument Panel #1), IC2 (Instrument Console #2), CC3 (Computer Console #3), etc., may be used when it is necessary to specify instrument or function location.

*** Normally inaccessible or behind-the-panel devices or functions may be depicted by using the same symbols but dashed horizontal bars, i.e.

Figure A-2. General Instrument or Function Symbols

First-Letters	Initiating or Measured Variable	Controllers Recording	Controllers Indicating	Controllers Blind	Self-Actuated Control Valves	Readout Recording	Readout Indicating	Switches High**	Switches Low	Switches Comb	Transmitters Recording	Transmitters Indicating	Transmitters Blind	Solenoids, Relays, Computing Devices	Primary Element	Test Point	Wall or Probe	Viewing Device, Glass	Safety Device	Final Element
A	Analysis	ARC	AIC	AC		AR	AI	ASH	ASL	ASHL	ART	AIT	AT	AY	AE	AP	AW			AV
B	Burner/Combustion	BRC	BIC	BC		BR	BI	BSH	BSL	BSHL	BRT	BIT	BT	BY	BE		BW	BG		BZ
C	User's Choice																			
D	User's Choice																			
E	Voltage	ERC	EIC	EC		ER	EI	ESH	ESL	ESHL	ERT	EIT	ET	EY	EE					EZ
F	Flow Rate	FRC	FIC	FC	FCV, FICV	FR	FI	FSH	FSL	FSHL	FRT	FIT	FT	FY	FE	FP		FG		FV
FQ	Flow Quantity	FQRC	FQIC	FQC		FQR	FQI	FQSH	FQSL			FQIT	FQT	FQY	FQE					FQV
FF	Flow Ratio	FFRC	FFIC	FFC		FFR	FFI	FFSH	FFSL			FFIT			FE					FFV
G	User's Choice																			
H	Hand		HIC	HC						HS										HV
I	Current	IRC	IIC	IC		IR	II	ISH	ISL	ISHL	IRT	IIT	IT	IY	IE					IZ
J	Power	JRC	JIC	JC		JR	JI	JSH	JSL	JSHL	JRT	JIT	JT	JY	JE					JV
K	Time	KRC	KIC	KC	KCV	KR	KI	KSH	KSL	KSHL	KRT	KIT	KT	KY	KE					KV
L	Level	LRC	LIC	LC	LCV	LR	LI	LSH	LSL	LSHL	LRT	LIT	LT	LY	LE		LW	LG		LV
M	User's Choice																			
N	User's Choice																			
O	User's Choice																			
P	Pressure, Vacuum	PRC	PIC	PC	PCV	PR	PI	PSH	PSL	PSHL	PRT	PIT	PT	PY	PE	PP			PSV, PSE	PV
PD	Pressure, Differential	PDRC	PDIC	PDC	PDCV	PDR	PDI	PDSH	PDSL		PDRT	PDIT	PDT	PDY	PE	PP				PDV
Q	Quantity	QRC	QIC	QC		QR	QI	QSH	QSL	QSHL	QRT	QIT	QT	QY	QE					QZ
R	Radiation	RRC	RIC	RC		RR	RI	RSH	RSL	RSHL	RRT	RIT	RT	RY	RE		RW			RZ
S	Speed Frequency	SRC	SIC	SC	SCV	SR	SI	SSH	SSL	SSHL	SRT	SIT	ST	SY	SE					SV
T	Temperature	TRC	TIC	TC	TCV	TR	TI	TSH	TSL	TSHL	TRT	TIT	TT	TY	TE	TP	TW		TSE	TV
TD	Temperature, Differential	TDRC	TDIC	TDC	TDCV	TDR	TDI	TDSH	TDSL		TDRT	TDIT	TDT	TDY	TE	TP	TW			TDV
U	Multivariable					UR	UI							UY						UV
V	Vibration, Machinery Analysis	VRC				VR	VI	VSH	VSL	VSHL	VRT	VIT	VT	VY	VE					VZ
W	Weight Force	WRC	WIC	WC	WCV	WR	WI	WSH	WSL	WSHL	WRT	WIT	WT	WY	WE					WZ
WD	Weight Force Differential	WDRC	WDIC	WDC	WDCV	WDR	WDI	WDSH	WDSL		WDRT	WDIT	WDT	WDY	WE					WDZ
X	Unclassified																			
Y	Even State Presence		YIC	YC		YR	YI	YSH	YSL				YT	YY	YE					YZ
Z	Position Dimension	ZRC	ZIC	ZC	ZCV	ZR	ZI	ZSH	ZSL	ZSHL	ZRT	ZIT	ZT	ZY	ZE					ZV
ZD	Gauging Deviation	ZDRC	ZDIC	ZDC	ZDCV	ZDR	ZDI	ZDSH	ZDSL		ZDRT	ZDIT	ZDT	ZDY	ZDE					ZDV

Note: This table is not all-inclusive.
Other Possible Combinations: FO - Restriction Orifice; FRK, HIK - Control Stations; FX - Accessories; TJR - Scanning Recorder; PFR - Ratio; LLH - Pilot Light; PFR - Ratio; KQI - Running Time Indicator; QQI - Indicating Counter; WKIC - Rate-of-Weight-Loss Controller; HMS - Hand Momentary Switch

* A, alarm, the annunciating device, may be used in the same fashion as S, switch, the actuating device.
** The letters H and L may be omitted in the undefined case.

Table A-3. Typical Letter Combinations

Appendix B: Glossary of Standard Process Instrumentation Terminology

APPENDIX B

Glossary of Standard Process Instrumentation Terminology

Note: The definitions presented in this appendix are mostly taken from the ANSI/ISA standard, "Process Instrumentation Terminology," S51.1-1979 (R 1993). This ANSI/ISA standard may be obtained from ISA. It contains many additional terms besides these, many notes and amplifying figures, and test procedures that relate to the terminology. The serious student of process control should have a copy of this standard available for routine reference. Bold-faced words refer to terms that are defined in separate entries.

accuracy—The degree of **conformity** of an indicated value to a recognized accepted standard value, or **ideal value**.

accuracy, measured—The maximum positive and negative **deviation** observed when testing a **device** under specified conditions and by a specified procedure.

accuracy rating—A number or quantity that defines a limit that errors will not exceed when a **device** is used under specified **operating conditions**.

actuating error signal—see **signal, actuating error**.

adaptive control—see **control, adaptive**.

adjustment span—A means provided in an instrument to change the slope of the input-output curve. See **span shift**.

adjustment, zero—A means provided in an instrument to produce a parallel shift of the input-output curve. See **zero shift**.

amplifier—A **device** that enables an **input signal** to control power from a source independent of the **signal** and thus be capable of delivering an output that bears some relationship to, and is generally greater than, the input signal.

analog signal—See **signal, analog**.

attenuation—(1) A decrease in **signal** magnitude between two points or between two frequencies. (2) The reciprocal of gain.

automatic control system—See **control system, automatic**.

automatic/manual station—A **device** that enables an operator to select an automatic **signal** or a manual signal as the input to a controlling element.

The automatic signal is normally the output of a **controller,** while the manual signal is the output of a manually operated device.

calibrate—To ascertain outputs of a **device** that correspond to a series of values of the quantity that the device is to measure, receive, or transmit. Data so obtained are used to do the following:

1.   determine the locations where scale graduations are to be placed;

2.   adjust the output, to bring it to the desired value, within a specified tolerance;

3.   ascertain the **error** by comparing the device output reading against a standard.

cascade control—See **control, cascade.**

characteristic curve—A graph (curve) that shows the ideal values at **steady state,** or an output variable of a system as a function of an input variable, where the other input variables are maintained at specified constant values. Note: When the other input variables are treated as **parameters,** a set of characteristic curves is obtained.

closed loop—see **loop, closed.**

closed-loop gain—see **gain, closed-loop.**

compensation—The provision of a special construction or , a supplemental **device,** circuit, or special materials to counteract sources of **error** caused by variations in specified **operating conditions.**

compensator—A **device** that converts a signal into some function that, either alone or in combination with other signals, directs the **final controlling element** to reduce deviations in the directly controlled variable.

conformity—Of a curve: the closeness to which it approximates a specified curve (such as logarithmic, parabolic, or cubic).

control action—Of a **controller** or a controlling system: the nature of the change of the output affected by the input. Note: The output may be a **signal** or a value of a **manipulated variable.** The input may be the control loop **feedback signal** when the set point is constant, an **actuating error signal,** or the output of another controller.

control action, derivative (rate) (D)—**Control action** where the output is proportional to the rate of change of the input.

control action, floating—Control action where the rate of change of the output variable is a predetermined function of the input variable. Note: The rate of change may have one absolute value, several absolute values, or any value between two predetermined values.

control action, integral (reset) (I)—Control action where the output is proportional to the time integral of the input; that is, the rate of change of output is proportional to the input.

control action, proportional (P)—Control action where there is a continuous linear relation between the output and the input. Note: This condition applies when both the output and input are within their normal operation ranges and when operation is at a frequency below a limiting value.

control action, proportional-plus-derivative (rate) (PD)—Control action where the output is proportional to a linear combination of the input and the time-rate-of-change of input.

control action, proportional-plus-integral (reset) (PI)—Control action where the output is proportional to a linear combination of the input and the time integral of the input.

control action, proportional-plus-integral (reset)-plus-derivative (rate) (PID)—Control action where the output is proportional to a linear combination of the input, the time integral of the input, and the time rate-of-change of input.

control, adaptive—Control where automatic means are used to change the type or influence (or both) of control **parameters** in such a way as to improve the performance of the **control system**.

control, cascade—Control where the output of one **controller** is the **set point** for another controller.

control center—An equipment structure, or group of structures, from which a **process** is measured, controlled, and/or monitored.

control, differential gap—Control where the output of a **controller** remains at a maximum or minimum value until the **controlled variable** crosses a band or gap, causing the output to reverse. The controlled variable must then cross the gap in the opposite direction before the output is restored to its original condition.

control, direct digital—Control performed by a digital **device** that establishes the **signal** to the **final controlling element**. Often called "DDC."

control, feedback—Control where a measured variable is compared to its **desired value** so as to produce an **actuating error signal** that is acted upon in such a way as to reduce the magnitude of the **error**.

control, feedforward—Control where information concerning one or more conditions that can disturb the **controlled variable** is converted, outside of any **feedback loop**, into corrective action so as to minimize deviations of the controlled variable.

control, high limiting—Control where the output **signal** is prevented from exceeding a predetermined high limiting value.

control, low limiting—Control where the output **signal** is prevented from exceeding a predetermined low limiting value.

control mode—A specific type of **control action** such as **proportional, integral,** or **derivative**.

control, optimizing—Control that automatically seeks and maintains the most advantageous value of a specified variable rather than maintaining it at one set value.

control, shared-time—Control where one **controller** divides its computation or control time among several control loops rather than by acting on all loops simultaneously.

control, supervisory—Control where the control loops operate independently, subject to intermittent corrective action, for example, **set point** changes from an external source.

control system—A system in which deliberate guidance or manipulation is used to achieve a prescribed value of a variable.

control system, automatic—A **control system** that operates without human intervention.

control system, multielement (multivariable) —A **control system** that utilizes **input signals** derived from two or more **process** variables for the purpose of jointly affecting the action of the control system.

control system, noninteracting—A **multielement control system** designed to avoid disturbances to other controlled variables caused by the **process** input adjustments that are made to control a particular process variable.

control, time proportioning—**Control** where the output **signal** consists of periodic pulses whose duration is varied so as to relate, in some prescribed manner, the time average of the output to the **actuating error signal**.

control valve—A **final controlling element** through which a fluid passes that adjusts the size of the flow passage as directed by a **signal** from a **controller** so as to modify the rate of flow of the fluid.

control, velocity limiting—Control in which the rate of change of a specified variable is prevented from exceeding a predetermined limit.

controlled system--See **system, controlled**.

controlled variable—See **variable, directly controlled**.

controller—A **device** that operates automatically to regulate a controlled variable.

controller, derivative (D)—A **controller** that produces **derivative control action** only.

controller, direct-acting—A **controller** in which the value of the **output signal** increases as the value of the input (measured variable) increases. See **controller, reverse-acting**.

controller, floating—A **controller** in which the rate of change of the output is a continuous (or at least a piecewise continuous) function of the **actuating error signal**.

controller, integral (reset) (I) —A **controller** that produces **integral control action** only.

controller, multiposition—A **controller** that has two or more discrete values of output.

controller, on-off—A **two-position controller** in which one of the two discrete values is zero.

controller, program—A **controller** that automatically holds or changes **set point** to follow a prescribed program for a **process**.

controller, proportional (P) —A **controller** that produces **proportional control action** only.

controller, proportional-plus-derivative (rate) (PD) —A **controller** that produces **proportional-plus-derivative (rate) control action**.

controller, proportional-plus-integral (reset) (PI)—A **controller** that produces **proportional-plus-integral (reset) control action**.

controller, proportional-plus-integral (reset) plus-derivative (rate) (PID) —A **controller** that produces **proportional-plus-integral (reset)-plus-derivative (rate) control action.**

controller, ratio—A **controller** that maintains a predetermined ratio between two variables.

controller, reverse-acting—A **controller** in which the value of the **output signal** decreases as the value of the input (measured variable) increases. See **controller, direct-acting.**

controller, sampling—A **controller** that uses intermittently observed values of a **signal,** such as the **set point signal** and the **actuating error signal** or the signal that represents the controlled variable, to effect **control action.**

controller, self-operated (regulator) —A **controller** in which all the energy to operate the **final controlling element** is derived from the **controlled system.**

controller, time schedule—A **controller** in which the **set point** or the **reference-input signal** automatically adheres to a predetermined time schedule.

controller, two-position—A **multiposition controller** that has two discrete values of output.

correction time—See **time, settling.**

damping—(1) (noun) The progressive reduction or suppression of oscillation in a **device** or system. (2) (adj.) Pertaining to or productive of damping.

dead band—The **range** through which an input can be varied without initiating observable response.

dead time—See **time, dead.**

dead zone—See **zone, dead.**

delay—The interval of time between a changing **signal** and its repetition for some specified duration at a downstream point of the signal path; the value θ in the transform factor exp $(-\theta s)$. See **time, dead.**

derivative action time—See **time, derivative action.**

derivative controller—See **controller, derivative (D).**

desired value—See **value, desired.**

detector—See **transducer.**

deviation—Any departure from a **desired value** or expected value or pattern.

deviation, steady-state—The system deviation after **transients** have expired.

deviation, system—The instantaneous value of the **directly controlled variable** minus the **set point.**

deviation, transient—The instantaneous value of the **directly controlled variable** minus its **steady-state** value.

device—An apparatus for performing a prescribed function.

differential gap control—See **control, differential gap.**

digital signal—See **signal, digital.**

direct-acting controller—See **controller, direct-acting.**

direct digital control—See **control, direct digital.**

directly controlled system—See **system, directly controlled.**

directly controlled variable—See **variable, directly controlled.**

distance/velocity lag—A delay attributable to the transport of material or to the finite rate of propagation of a **signal.**

disturbance—An undesired change that takes place in a **process** that tends to affect adversely the value of a controlled variable.

drift—An undesired change in the input-output relationship over a period of time.

drift, point—The change in output over a specified period of time for a constant input under specified reference operating conditions.

droop—See **offset.**

dynamic gain—See **gain, dynamic.**

dynamic response—See **response, dynamic.**

element—A component of a **device** or system.

element, final controlling—The **forward controlling element** that directly changes the value of the **manipulated variable.**

element, primary—The system **element** that quantitatively converts the **measured variable** energy into a form suitable for measurement.

element, reference-input—The portion of the **controlling system** that changes the reference-input **signal** in response to the **set point**.

element, sensing—The **element** that is directly responsive to the value of the **measured variable**.

elements, feedback—Those **elements** in the **controlling system** that act to change the **feedback signal** in response to the **directly controlled variable**.

elements, forward controlling—Those **elements** in the **controlling system** that act to change a variable in response to the actuating **error signal**.

error—The algebraic difference between the indication and the **ideal value** of the **measured signal**. It is the quantity that, algebraically subtracted from the indication, gives the ideal value.

error signal—See **signal, error.**

error, systematic—An **error** that, in the course of making a number of measurements under the same conditions of the same value of a given quantity, either remains constant in absolute value and sign or varies according to a definite law when the conditions change.

excitation—The external supply that is applied to a **device** so it operates properly.

feedback control—See **control, feedback.**

feedback elements—See **elements, feedback.**

feedback loop—See **loop, closed (feedback loop).**

feedback signal—See **signal, feedback.**

feedforward control—See **control, feedforward.**

final controlling element—See **element, final controlling.**

floating control action—See **control action, floating.**

floating controller—See **controller, floating.**

flowmeter—A **device** that measures the rate of flow or quantity of a moving fluid in an open or closed conduit. It usually consists of both a primary and a secondary device.

frequency response characteristic—The frequency-dependent relation, in both amplitude and phase, between **steady-state** sinusoidal inputs and the resulting fundamental sinusoidal outputs.

gain, closed-loop—The **gain** of a **closed-loop** system, expressed as the ratio of the output change to the input change at a specified frequency.

gain, dynamic—The magnitude ratio of the **steady-state** amplitude of the **output signal** from an **element** or **system** to the amplitude of the **input signal** to that element or system for a sinusoidal **signal.**

gain, loop—The ratio of the change in the **return signal** to the change in its corresponding **error signal** at a specified frequency.

gain, open-loop—See **gain, loop.**

gain, proportional—The ratio of the change in output caused by **proportional control action** to the change of the input.

gain, static (zero-frequency gain) —The **gain** of an **element**, or **loop gain** of a system; the value approached as a limit as frequency approaches zero.

hardware—Physical equipment that is directly involved in performing industrial **process** measuring and controlling functions.

hunting—An undesirable oscillation of appreciable magnitude that is prolonged after external stimuli disappear.

hysteresis—That property of an **element** that is evidenced by the dependence of the value of the output, for a given excursion of the input, upon the history of prior excursions and the direction of the current traverse.

I controller—See **controller, integral (reset) (I).**

ideal value—See **value, ideal.**

indicating instrument—See **instrument, indicating.**

inherent regulation—See **self-regulation.**

input—See **signal, input.**

instrument, computing—A **device** in which the output is related to the input or inputs by a mathematical function such as addition, averaging, division, integration, lead-lag, signal limiting, squaring, square root extraction, or subtraction.

instrument, indicating—A measuring instrument in which only the present value of the measured variable is visually indicated.

instrument, measuring—A **device** for ascertaining the magnitude of a quantity or condition presented to it.

instrument, recording—A measuring instrument in which the values of the measured variable are recorded.

instrumentation—A collection of instruments or their application for the purpose of observing, measuring, or controlling.

integral action limiter—A **device** that limits the value of the **output signal** due to **integral control action** to a predetermined value.

integral action rate (reset rate) —(1) Of **proportional-plus-integral** or **proportional-plus-integral-plus-derivative control action devices**: for a step input, the ratio of the initial rate of change of output due to **integral control action** to the change in **steady-state** output due to **proportional control action.** Note: Integral action rate is often expressed as the number of repeats per minute because it is equal to the number of times per minute that the proportional response to a step input is repeated by the initial integral response. (2.) Of integral control action devices: for a step input, the ratio of the initial rate of change of output to the input change.

integral control action—See **control action, integral**.

integral controller—See **controller, integral (reset)**.

interference, common mode—A form of interference that appears between measuring circuit terminals and ground.

interference, differential mode—See **interference, normal mode**.

interference, normal mode—A form of interference that appears between measuring circuit terminals.

intrinsically safe equipment and wiring—Equipment and wiring that under normal or abnormal conditions are incapable of releasing sufficient electrical or thermal energy to cause ignition of a specific hazardous atmospheric mixture in its most easily ignited concentration.**linear system**. See **system, linear.**

linearity—The degree to which a curve approximates a straight line.

load regulation—The change in output (usually speed or voltage) from no-load to full-load (or other specified load limits). See **offset**.

loop, closed (feedback loop)—A **signal** path that includes a forward path, a **feedback** path, and a **summing point** and that forms a closed circuit.

loop, feedback—See **loop, closed (feedback loop)**.

loop gain—See **gain, loop**.

loop, open—A **signal** path without **feedback**.

loop transfer function—Of a **closed loop**: the **transfer function** obtained by taking the ratio of the **Laplace transform** of the **return signal** to the Laplace transform of its corresponding **error signal**.

manipulated variable—See **variable, manipulated**.

modulation—The process, or result of the process, whereby some characteristic of one wave is varied in accordance with some characteristic of another wave.

module—An assembly of interconnected components that constitutes an identifiable **device**, instrument, or piece of equipment. A module can be disconnected, removed as a unit, replaced with a spare. It has definable performance characteristics that permit it to be tested as a unit.

multielement control system—See **control system, multielement (multivariable)**.

multiposition controller—See **controller, multiposition**.

multivariable control system—See **control system, multielement (multivariable)**.

noise—An unwanted component of a **signal** or variable.

noninteracting control system—See **control system, noninteracting**.

normal mode rejection—The ability of a circuit to discriminate against a **normal mode voltage**.

normal mode voltage—See **voltage, normal mode**.

offset—The **steady-state deviation** when the **set point** is fixed. See also **deviation, steady-state**.

on-off controller—See **controller, on-off**.

operating conditions—The conditions to which a **device** is subjected, but not including the variable measured by the device.

optimizing control—See **control, optimizing**.

output signal—See **signal, output**.

overdamped—See **damping**.

overshoot—See **transient overshoot**.

parameter—A quantity or property that is treated as a constant but that may sometimes vary or be adjusted.

P controller—See **controller, proportional**.

PD controller—See **controller, proportional-plus-derivative**.

PI controller—See **controller, proportional-plus-integral**.

PID controller—See **controller, proportional-plus-integral-plus-derivative**.

position—Of a **multiposition controller**: a discrete value of the **output signal**.

primary element—See **element, primary**.

process—The physical or chemical change of matter or conversion of energy, such as, change in pressure, temperature, speed, or electrical potential.

process control—The regulation or manipulation of the variables that influence the conduct of a **process** in such a way as to obtain a product of desired quality and quantity in an efficient manner.

process measurement—The acquisition of information that establishes the magnitude of **process** quantities.

proportional band—The change in input required to produce a full-range change in output due to **proportional control action**.

proportional gain—See **gain, proportional**.

range—The region between the limits within which a quantity is measured, received, or transmitted; expressed by stating the lower and upper range values.

rate—See **control action, derivative**.

ratio controller—See **controller, ratio.**

reference-input element—See **element, reference-input.**

regulator—See **controller, self-operated (regulator).**

reliability—The probability that a device will perform its objective adequately, for the period of time specified under the **operating conditions** specified.

repeatability—The closeness of agreement among a number of consecutive measurements of the output for the same value of the input under the same **operating conditions**, approaching from the same direction, for full **range** traverses. It does not include hysteresis.

reproducibility—The closeness of agreement among repeated measurements of the output for the same value of input made under the same **operating conditions** over a period of time, approaching from both directions. It includes **hysteresis, dead band, drift,** and **repeatability.**

reset control action—See **control action, integral (reset).**

reset rate—See **integral action rate.**

resolution—The least interval between two adjacent discrete details that can be distinguished one from the other.

resonance—Of a system or **element**: a condition evidenced by large oscillatory amplitude that results when a small amplitude of periodic input has a frequency that approaches one of the natural frequencies of the driven system.

response, dynamic—The behavior of the output of a **device** as a function of the input, both with respect to time.

response, ramp—The total (transient plus **steady-state**) **time response** that results from a sudden increase in the rate of change from zero to some finite value of the input stimulus.

response, step—The total (transient plus **steady-state**) **time response** that results from a sudden change from one constant level of input to another.

response, time—An output, expressed as a function of time, that results from the application of a specified input under specified operating conditions.

reverse-acting controller—See **controller, reverse-acting.**

rise time—See **time, rise.**

sampling controller—See **controller, sampling**.

sampling period—The time interval between observations in a periodically sampling **control system**.

scale factor—The factor by which the number of scale divisions indicated or recorded by an instrument should be multiplied to compute the value of the **measured variable**.

self-operated controller—See **controller, self-operated (regulator)**.

self-regulation (inherent regulation) —The property of a **process** or machine that permits equilibrium to be attained after a **disturbance** without the intervention of a **controller**.

sensing element—See **element, sensing**.

sensitivity—The ratio of the change in output magnitude to the change of the input that causes it after the **steady state** has been reached.

sensor—See **transducer**.

servomechanism—An automatic **feedback control device** in which the controlled variable is mechanical position or any of its time derivatives.

set point—An input variable that sets the desired value of the controlled variable.

settling time—See **time, settling**.

shared-time control—See **control, shared-time**.

signal—A physical variable; one or more **parameters** that carry information about another variable (that the signal represents).

signal, actuating error—The **reference-input signal** minus the **feedback signal**.

signal, analog—A **signal** representing a variable that may be continuously observed and continuously represented.

signal converter—See **signal transducer**.

signal, digital—The representation of information by a set of discrete values in accordance with a prescribed law. Numbers represent these values.

signal, error—In a **closed loop**, the **signal** that results from subtracting a particular **return signal** from its corresponding **input signal**.

signal, feedback—The return signal that results from a measurement of the directly **controlled variable**.

signal, feedforward—See **control, feedforward**.

signal, input—A **signal** applied to a **device, element,** or **system**.

signal, measured—The electrical, mechanical, pneumatic, or other variable applied to the input of a **device**. It is the analog of the **measured variable** that is produced by a **transducer** (when such is used.)

signal, output—A **signal** delivered by a **device, element,** or **system**.

signal, reference-input—One external to a control loop, which serves as the standard of comparison for the **directly controlled variable**.

signal, return—In a closed loop, the results from a particular **input signal,** and transmitted by the loop, and to be subtracted from the **input signal**. See also **signal, feedback**.

signal selector—A **device** that automatically selects either the highest or the lowest input **signal** from among two or more **input signals**.

signal-to-noise ratio—The ratio of **signal** amplitude to **noise** amplitude.

signal transducer (signal converter) —A **transducer** that converts one standardized transmission **signal** to another.

span—The algebraic difference between the upper- and lower-range values.

span shift—Any change in slope of the input-output curve.

static gain—See **gain, static**.

steady state—A characteristic of a condition, such as a value, rate, periodicity, or amplitude, that exhibits only negligible change over an arbitrarily long period of time.

steady-state deviation—See **deviation, steady-state**.

step response—See **response, step**.

step-response time—See **time, step-response**.

summing point—Any point at which **signals** are added algebraically.

supervisory control—See **control, supervisory**.

system, control—See **control system**.

system, controlled—The collective functions performed in and by the equipment in which the variable(s) is (are) to be controlled.

system, controlling—(1) Of a feedback control system: that portion that compares the functions of a **directly controlled variable** and a **set point** and adjusts a **manipulated variable** as a function of the difference. It includes the **reference-input elements, summing point, forward** and **final controlling elements,** and **feedback elements** (including **sensing element**). (2) Of a **control system** without **feedback**: that portion that manipulates the **controlled system.**

system, directly controlled—The body, **process,** or machine that is directly guided or restrained by the **final controlling element** so as to achieve a prescribed value of the **directly controlled variable.**

system, indirectly controlled—The portion of the **controlled system** where the indirectly controlled variable is changed in response to changes in the **directly controlled variable.**

system, linear—A system whose **time response** to several simultaneous inputs is the sum of those inputs' independent **time responses.**

time constant—The value t in an exponential response term. Note: For the output of a first-order system that is forced by a step or an impulse, τ is the time required to complete 63.2 percent of the total rise or decay, at any instant during the process. τ is the quotient of the instantaneous ratio of change divided into the change still to be completed. In higher-order systems, there is a time constant for each of the first-order components of the process.

time, correction—See **time, settling.**

time, dead—The interval of time between the initiation of an input change or stimulus and the start of the resulting observable response.

time, derivative action—In **proportional-plus-derivative control action** for a unit ramp **signal** input, the advance in time of the **output signal** (after **transients** have subsided) caused by **derivative control action,** as compared to the output signal caused by **proportional control action** only.

time proportioning control—See **control, time proportioning.**

time response—See **response, time.**

time, rise—The time required for the output of a system (other than first-order) to change from a small specified percentage (often 5 or 10) of the **steady-state** increment to a large specified percentage (often 90 to 95), either before or in the absence of overshoot.

time, settling—The time required, after the initiation of a specified stimulus to a system, for the output to enter and remain with a specified narrow band centered on its **steady-state** value.

time, step-response—Of a **system** or an **element**: the time required for an output to change from an initial value to a large specified percentage of the final **steady-state** value either before or in the absence of overshoot, as a result of a step change to the input.

transducer—An **element** or **device** that receives information in the form of one quantity and converts it to information in the form of the same or another quantity.

transfer function—A mathematical, graphical, or tabular statement of the influence that a **system** or **element** has on a **signal** or action that is being compared at input and at output terminals.

transient—The behavior of a variable during transition between two **steady states**.

transient overshoot—The maximum excursion beyond the final **steady-state** value of output as the result of an input change.

transmitter—A **transducer** that responds to a **measured variable** by means of a sensing element and converts it to a standardized transmission **signal**, which is a function only of the **measured variable**.

value, desired—The value of the **controlled variable** that the user wants or chooses.

value, ideal—The value of the indication, output, or ultimately controlled variable of an idealized **device** or system.

variable, directly controlled—In a control loop, the variable that is sensed so as to originate a **feedback signal**.

variable, indirectly controller—A variable that does not originate a **feedback signal** but that is related to, and influenced by, the **directly controlled variable**.

variable, manipulated—A quantity or condition that is varied as a function of the **actuating error signal** so as to change the value of the **directly controlled variable**.

velocity limit—A limit that the rate of change of a specified variable may not exceed.

velocity limiting control—See **control, velocity limiting**.

voltage, common mode—A voltage of the same polarity on both sides of a differential input relative to ground.

voltage, normal mode—A voltage induced across the input terminals of a **device**.

zero shift—Any parallel shift of the input-output curve.

zone, dead—(1) For a **multiposition controller**, a zone of input where no value of the output exists. It is usually intentional and adjustable. (2) The term *dead zone* is sometimes used to denote **dead band**.

Appendix C:
Solutions to
All Exercises

APPENDIX C

Solutions to All Exercises

Unit 2

Exercise 2-1.

Controlled variable: oven temperature
Manipulated variable: electric current
Disturbances:
- ambient temperature
- oven contents
- endothermic and exothermic reactions
- leaks
- door opening and closing

Exercise 2-2.

Controlled variable: water temperature
Manipulated variable: gas flow
Disturbances:
- water usage
- Btu content of gas
- ambient temperature
- inlet water temperature

Exercise 2-3.

Manual system:
    1. Take a sample of pool water.
    2. Use litmus paper or equivalent to measure pH.
    3. Add acidic solution as necessary.

For this manual system--
    Controlled variable: pool water pH
    Manipulated variable: acidic solution added
Disturbances:
- variation in strength of acidic solution
- evaporation rate
- pH of any makeup water added to pool

Exercise 2-4.

A feedback automatic control system might appear as follows:

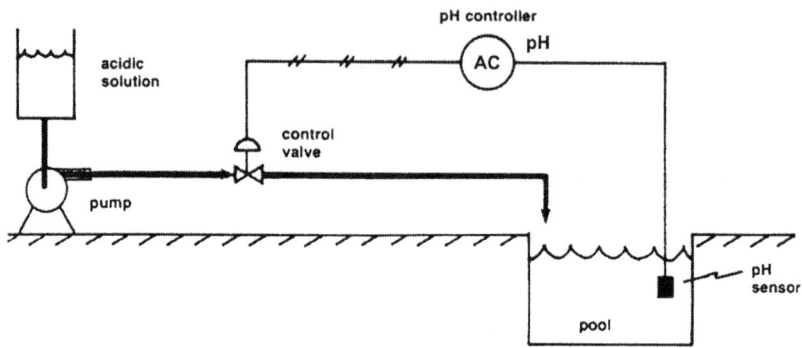

Exercise 2-5.

Set point: desired speed
Controlled variable: actual speed of vehicle
Manipulated variable: accelerator
Disturbances:
- grade of road
- weight
- engine performance

Note: For the technical purist, the intake manifold pressure is measured and controlled.

Exercise 2-6.

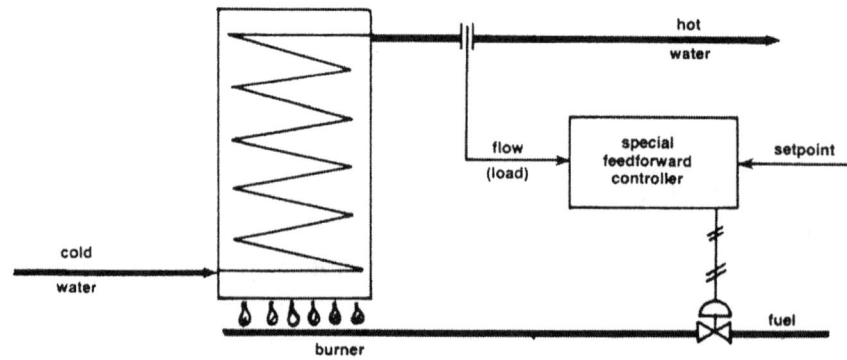

Exercise 2-7.

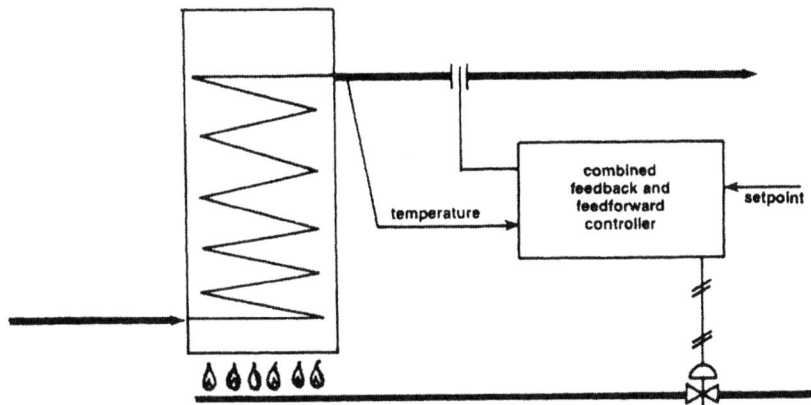

Exercise 2-8.

A moon shot is based on a predetermination of flight path (trajectory). To arrive at a specific point in space on the surface of the moon, NASA's calculations involve considering the gravitational effects on all bodies, the motions of all bodies, the rocket thrust, and the rocket altitude. NASA uses an elaborate computer model to determine fight trajectory.

A mid-course correction is needed to correct for inadequacies in the model and in the less-than-perfect performance of all the hardware components of the rocket system. The mid-course correction is a feedback adjustment to eliminate the errors between the desired and actual flight trajectory.

Exercise 2-9.

Consider, for example, a small home thermostat that allows the input of a normal maximum desired temperature, for example, 78°F, and the input of a normal minimum desired temperature, for example, 68°F. Various wider limits can be in effect during the evening hours while the occupants are asleep, for example, 80°F and 65°F. Also, during the day when the occupants are away at work or away on a trip, wider bands may also be programmed. A built-in clock can be included in the thermostat case so that room temperature can be brought back within normal control limits at prescribed times. This type of management of home temperature can be programmed on a routine seven-day week clock, and various subroutines can be adjusted as necessary to reflect the travel and living patterns of the occupants. Of course, one portion of the thermostat's operation is to activate the necessary heating or air conditioning systems when the actual temperature ranges outside of the desired band width.

Unit 3

Exercise 3-1.

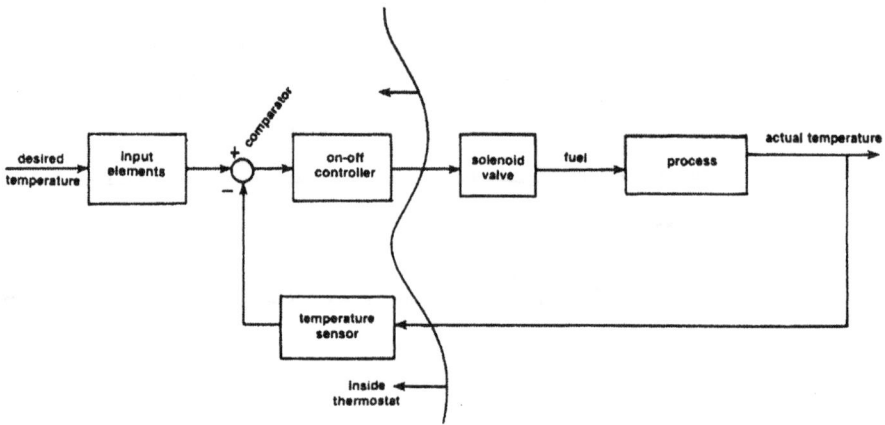

Exercise 3-2.

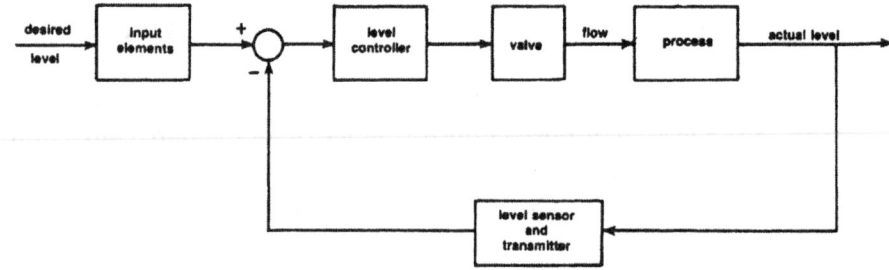

Exercise 3-3.

$$\frac{12 \text{ psi}}{200°F} = 0.06 \text{ psi per } °F$$

Exercise 3-4.

$$\frac{16 \text{ mA}}{60 \text{ gpm}} = 0.266 \text{ mA per gpm}$$

Exercise 3-5.

Input elements: no
Electronic transmission: no
Large pneumatic valve: probably yes
2,000 ft. pneumatic transmission system: yes
Pneumatic controller: no
Electronic controller: no
Bare thermocouple: no
Orifice meter: no
Chromatograph: yes, probably
Thermocouple in thermowell: yes, probably

Exercise 3-6.

Liquid flow: very rapid
Gaseous flow: less rapid than liquid flow
Liquid level in a small tank: moderately rapid
Liquid pressure in an enclosed tank: rapid
Gaseous pressure in a large tank: slow
Composition in a large distillation column: slow
Temperature in a liquid-filled tank: slow
Your coworkers: you be the judge!

Unit 4

Exercise 4-1.

The distinctions between transmission to a board-mounted (central control room) controller and to a board indicator and/or recorder are minor. The transmission systems are virtually identical. The only difference is in the usage of the signal once it gets to the control room.

Exercise 4-2.

The thermowell causes more lag and slows down the temperature response. The curve in Fig. 4-2 might be more S-shaped if a thermowell were used:

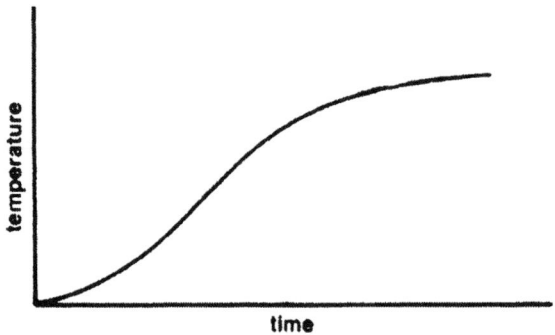

Exercise 4-3.

If all three of the following graphs are superimposed on one another (to a rough approximation), the use of the sampling will introduce a dead time that is roughly equivalent to one-half of the sampling interval T.

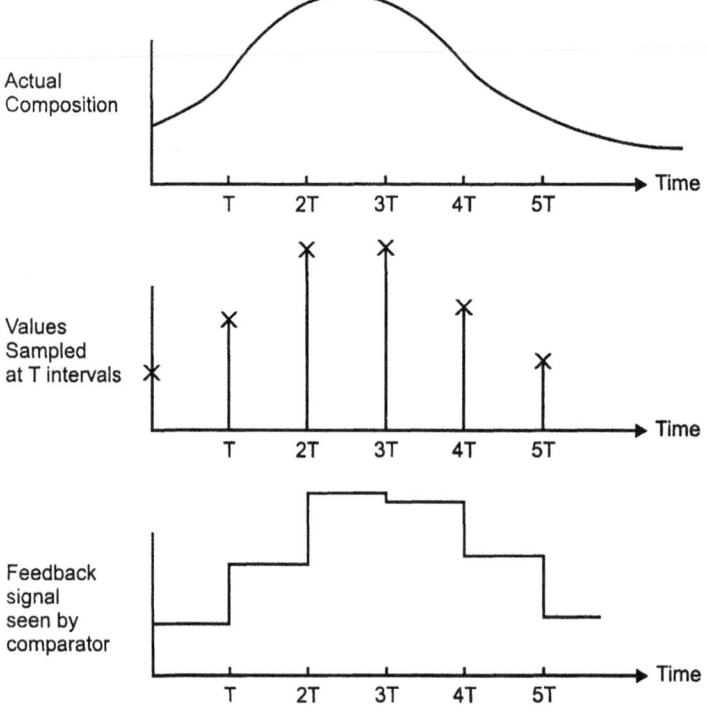

Exercise 4-4.

If there are errors in accuracy, they will not introduce any instability or disturbance into the process, but they may cause the operator to need to readjust his or her set point. This is a minor irritant, but not nearly so serious as instabilities caused by lack of precision or reproducibility.

Exercise 4-5.

Refer to Fig. 4-8. A time constant of three seconds dictates a maximum pneumatic transmission distance of 700 ft.

Exercise 4-6.

Both transmission systems are of equivalent importance insofar as their dynamics are concerned.

Exercise 4-7.

    a.   No systematic error of significance. Apparent low precision error.

    b.   Precision high, accuracy acceptable. No significant systematic error.

    c.   Precision poor. Systematic aim to the right.

    d.   Low precision error. Significant bias (an archer can see, but an experimenter never can).

Unit 5

Exercise 5-1.

Pressure = force/area. A cubic foot of water is 62.4 lb spread over 144 square inches. A foot of water is equal to the following:

$$\frac{62.4}{144} = 0.433 \text{ psi}$$

Thirty inches of water is therefore equal to the following:

$$0.433 \times \frac{30}{12} = 1.08 \text{ psi gage}$$

The high pressure is therefore:

$$14.6 + 1.08 = 15.68 \text{ psig}$$

Exercise 5-2.

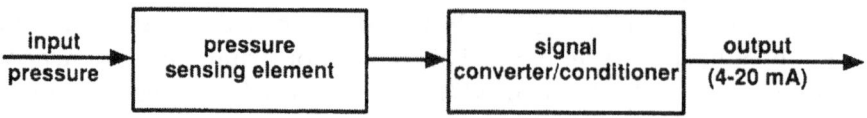

Exercise 5-3.

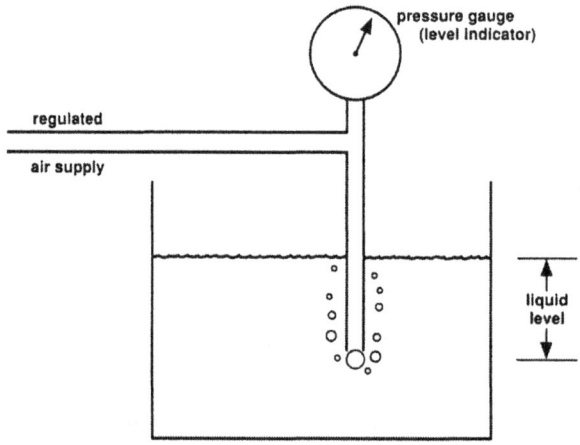

Exercise 5-4.

As the level rises, the apparent weight of the displacer decreases, thereby showing a proportional and linear relationship between weight and level.

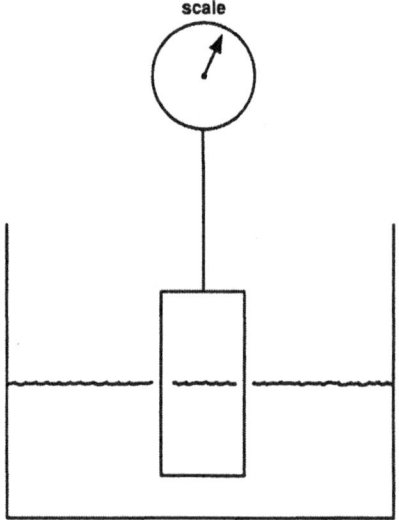

Exercise 5-5.

$$v = 0.6 \sqrt{(2)(32.2 \text{ ft/sec}^2)\left(\frac{16}{12} \text{ ft}\right)} = 5.56 \text{ ft/sec}$$

Exercise 5-6.

A turbine flowmeter has a meter coefficient defined as follows:

$$K = \frac{\text{cycles per time}}{\text{volume per time}} = \frac{\text{cycles}}{\text{volume}} = \text{meter coefficient}$$

Flow rates can be converted to total flow by a totalizer-type instrument that simply counts the total cycles.

Exercise 5-7.

$$\text{Voltage} = (38 \ \mu V/°C)(80°C - 50°C)(1 \ mV/1{,}000\mu V) = 1.14 \ mV$$

Exercise 5-8.

The thermistor shows a negative temperature coefficient and limited temperature range. The RTD curve has the least nonlinearity and a broad temperature range. The curve for the thermocouple shows the broadest temperature range, the greatest nonlinearity, and a region of very low sensitivity. Therefore, the RTD is the most stable, the thermocouple is the most versatile, and the thermistor is the most sensitive.

Unit 6

Exercise 6-1.

    a.   0.5

    b.   500 percent

Exercise 6-2.

At steady state, there is an unchanging error. This error produces a controller output that produces an unchanging amount of manipulated variable. All of the manipulated variable so produced is used to maintain the current state, and there is no "surplus" or incremental amount of the manipulated variable available to produce a change from the current steady state. In effect, in maintaining the status quo the manipulated variable output is consumed.

Exercise 6-3.

It makes no difference whatsoever if the error is caused by a change in disturbance. The dynamics of the response will be similar either way. Of course, the transient response of the process will lead to a new final value if there has been a change in set point.

Exercise 6-4.

If manual reset is used to bias the controller output to eliminate offset at a specific operating point, there will be no offset in evidence until there is either a change to a new set point or there is a change in some of the disturbances at work on the system. Since disturbances and set points change frequently, eliminating offset by manual reset is not a long-term operating solution.

Exercise 6-5.

The proposed controller would have good dynamic response for large errors, and when the error is small the reset-only controller could be used to trim out offset. The significant advantage would be that the dynamic lag penalties inherent with reset would not be present except when that mode is in active use.

Exercise 6-6.

The two algorithms are equivalent, but the noninteracting form is a bit easier to tune. Because all of the modes are independent of each other, when you change the gain of one mode it does not produce a change in the gain of the other modes.

Exercise 6-7.

Filtering is necessary on the rate mode because otherwise very small—but rapid—changes in the error signal might produce very large outputs from the rate mode. This would be deleterious to loop dynamics.

Exercise 6-8.

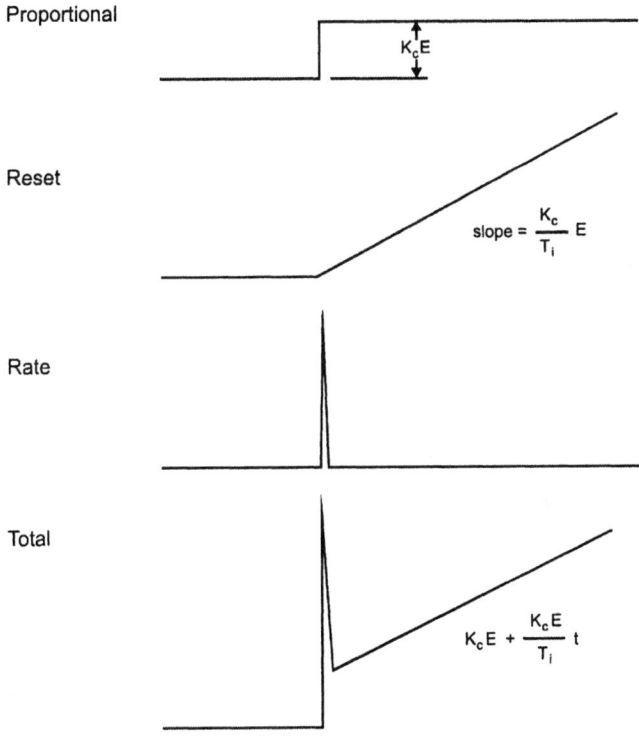

Exercise 6-9.

Proportional

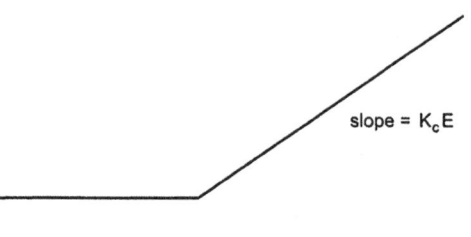

slope = $K_c E$

Reset

parabolic shape

Rate

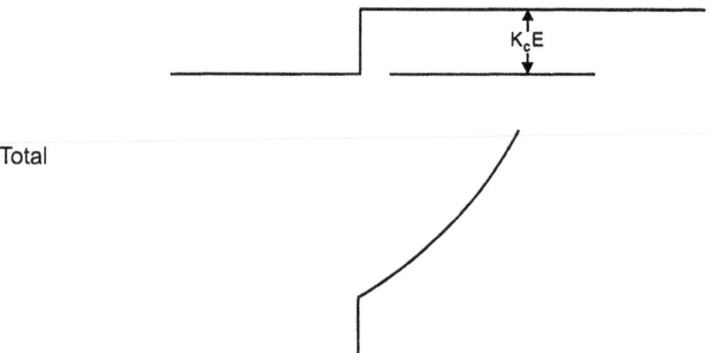

$K_c E$

Total

Unit 7

Exercise 7-1.

A 0.5 psig air-to-valve signal increase produces a flow increase of 15 percent. Therefore, at a base flow of 20 gpm, the increase is 3 gpm.

Exercise 7-2.

For a linear valve, the gain of the valve does not change. Therefore, at 20 gpm a 0.5 psi increase produces a change of 1.5 gpm.

Exercise 7-3.

$$\text{Flow, } M_L = C_v \sqrt{\frac{\Delta P}{G}}$$

$$= 20 \sqrt{\frac{10}{1}}$$

$$= (20)(3.16)$$

$$= 63.3 \text{ gpm}$$

Exercise 7-4.

$$\text{Flow, } M_g = 63.3 C_v Y \sqrt{\Delta P Y_1}$$
$$= (63.3)(32)(0.8) \sqrt{(10)(3.103)}$$
$$= (1620) \sqrt{31.03}$$
$$= (1620)(5.57)$$
$$= 9,027 \text{ lbs per hour}$$

Exercise 7-5.

Refer to Fig. 7-8. With 10 percent of the dynamic drop across the value and the valve operating in the range of 50 to 70 percent open, the valve characteristic curve is effectively linear.

Exercise 7-6.

For most situations, we prefer the valve to fail closed if it is adding energy to the system (e.g., a steam valve) but to fail open if it is removing energy from the system (e.g., a cooling water valve).

Unit 8

Exercise 8-1.

Curve 1: Time constant of approximately 1 minute
Curve 2: Time constant of approximately 3.5 minutes
Curve 3: Time constant of approximately 7.5 minutes

Exercise 8-2.

A time constant of zero implies instantaneous correspondence of output to input. The output is a straight multiple of the input with no time-delayed dynamics. A time constant of infinity implies no response whatsoever.

Exercise 8-3.

Time constant = capacitance/conductance = 28 ft^3/6 ft^3 per min = 4.66 min

Exercise 8-4.

Time constant = capacitance/conductance
= mass * specific heat / (area * film coefficient)
= 0.26 lb * 0.6 Btu/(lb*^{o}F) / (0.16 ft^2 * 4.4 Btu/(min*ft^2*^{o}F))
= 0.22 min

Exercise 8-5.

Dead time = 3.6

Exercise 8-6.

$$Dead\ time\ =\ distance/velocity$$
$$=\ \frac{16\ ft}{21\ ft/min}\ =\ 0.762\ min$$

Exercise 8-7.

As the controller gain is increased, the response of the loop becomes more dynamic and much more rapid; there is an increase in the speed of response. This increased speed tends to make the loop less stable. Finding the optimum controller gain implies making a trade-off between increased speed of response and decreased stability.

Exercise 8-8.

As more and more dead time is introduced, the loop becomes less and less stable. During the dead time of a component, there is no output, and thus no corrective action is initiated.

Unit 9

Exercise 9-1.

For two-mode and three-mode controllers, there can be many response curves with a decay ratio of one-quarter but with oscillations at different frequencies.

Exercise 9-2.

Tuning parameters are not a function of whether loop type, such as a liquid-level loop or a flow loop. Rather, tuning constants depend upon the individual dynamic parameters and the individual gains (or sensitivities) of the various components that make up the loop. Since these vary from one loop to the next, the desirable tuning parameters also vary from loop to loop.

Exercise 9-3.

Proportional only:

$$S_c = 0.5\, S_u$$

$$= (0.5)\,(0.3\ \text{psi/ft})$$

$$= 0.15\ \text{psi/ft}$$

Proportional-plus-reset:

$$S_c = 0.45\, Su$$

$$= (0.45)\,(0.3\ \text{psi/ft})$$

$$= 0.135\ \text{psi/ft}$$

$$T_i = P_u\,/\,1.2 = 3\,/\,1.2 = 2.5\ \text{min}$$

Proportional-plus-derivative:

$$S_c = 0.6\ S_u$$

$$= (0.6)\ (0.3)$$

$$= 0.18\ \text{psi/ft}$$

$$T_d = P_u / 8$$

$$= 0.375\ \text{min}$$

Proportional-plus-reset-plus-derivative:

$$S_c = 0.6\ S_u$$

$$= 0.18\ \text{psi/ft}$$

$$T_i = 0.5\ P_u = 1.5\ \text{min}$$

$$T_d = P_u / 8 = 0.375\ \text{min}$$

Exercise 9-4.

The general units of Sp are the units of the process output divided by the process input. The product of all of the sensitivities around the feedback loop, $SpSc$, is without units or dimensions.

Exercise 9-5.

$$R_1 = R_r * S_t / \text{Step Size}$$

$$= 0.6\ (\text{ft/(psi} * \text{min)}) * 2\ (\text{psi/ft}) / 1.2\ (\text{psi})$$

$$= 1.0 / \text{min}$$

Proportional:

$$K_c = 1 / L_r R_1 = 1 / ((0.15\ \text{min}\ ) * (1.0 / \text{min}\ ))$$

$$= 6.7\ (\text{dimensionless})$$

Proportional-plus-reset:

$$K_c = 0.9 \ / \ L_r R_1 = 0.9 \ / \ ((\ 0.15 \ \text{min}) * (\ 1.0 \ / \ \text{min}))$$

$$= 6.0$$

$$T_i = 3.33 \ L_r = (3.33) * (0.15 \ \text{min})$$

$$= 0.50 \ \text{min}$$

Proportional-plus-reset-plus-rate:

$$K_c = 1.2 \ / \ L_r R_1 = 1.2 \ / \ ((0.15 \ \text{min} \ ) * (1.0 \ / \ \text{min}))$$

$$= 8.0$$

$$T_i = 2.0 \ L_r = 2 * 0.15 \ \text{min} = 0.30 \ \text{min}$$

$$T_d = 0.5 \ L_r = 0.5 * 0.15 \ \text{min} = 0.075 \ \text{min}$$

Exercise 9-6.

Proportional:

$$PB = 100 \ / \ K_c$$

$$= 100 \ / \ 6.7 \ = 15\%$$

Proportional-plus-reset:

$$PB = 100 \ / \ K_c$$

$$= 100 \ / \ 6 = 16.7\%$$

$$\text{Reset Rate} = 1 \ / \ T_i$$

$$= 1 \ / \ 0.50 \ \text{min}$$

$$= 2 \ \text{repeats per minute}$$

Proportional-plus-reset-plus-rate:

$$PB = 100 / K_c = 100 / 8 = 12.5\%$$

$$\text{Reset Rate} = 1 / T_i$$

$$= 1 / 0.30 \text{ min}$$

$$= 3.3 \text{ repeats per minute}$$

$$T_d = 0.075 \text{ min}$$

Unit 10

Exercise 10-1.

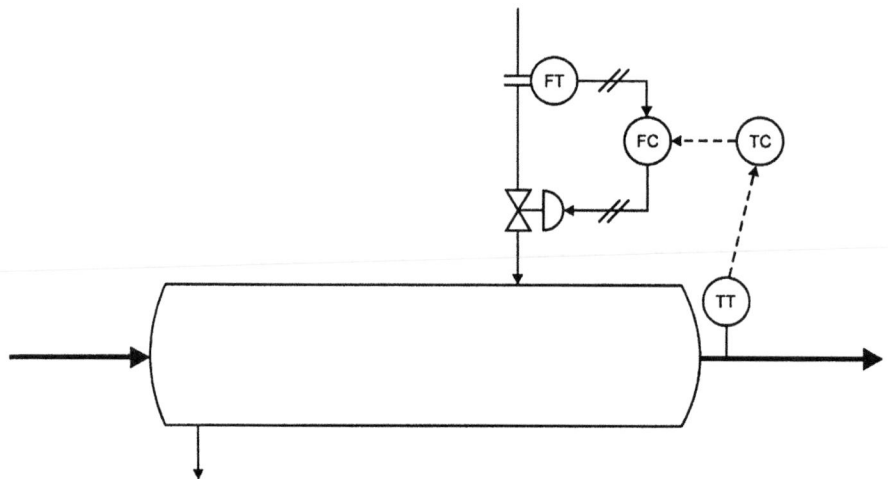

Exercise 10-2.

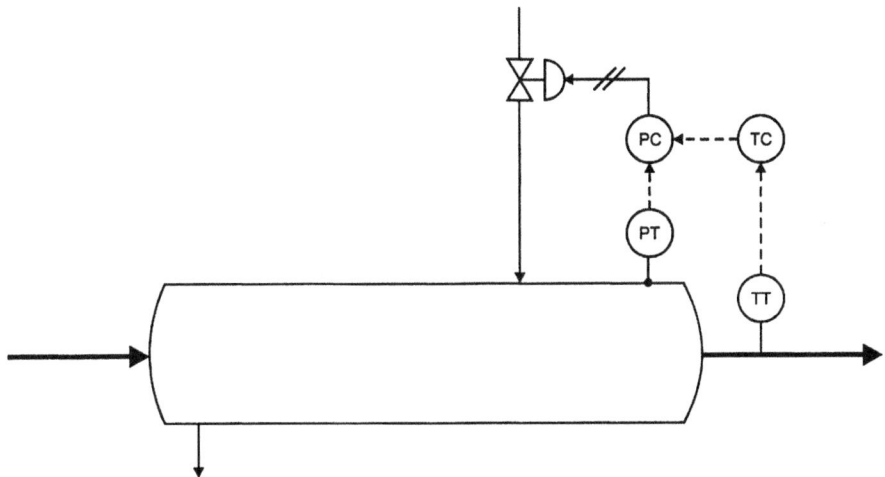

Exercise 10-3.

The tubes will be practically at the temperature that corresponds to the steam-side pressure because the liquid film offers a much greater resistance to heat transfer. The TC-PC is faster.

Exercise 10-4.

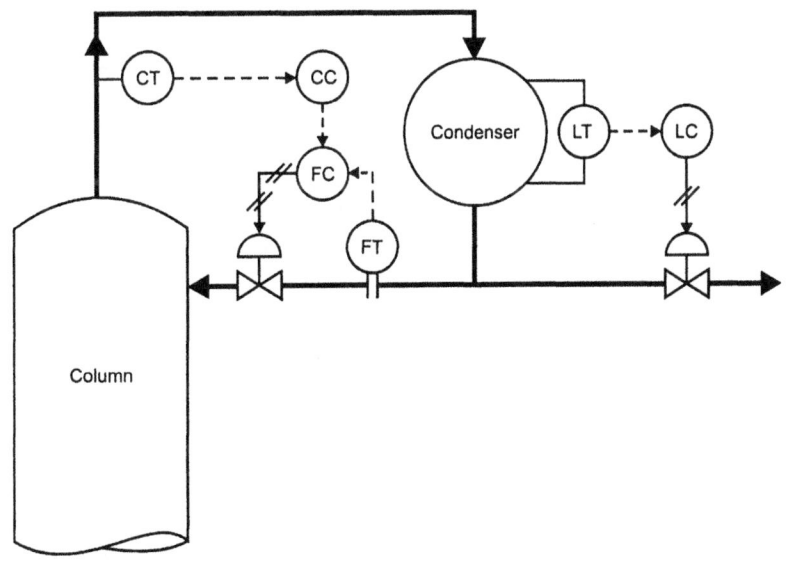

Exercise 10-5.

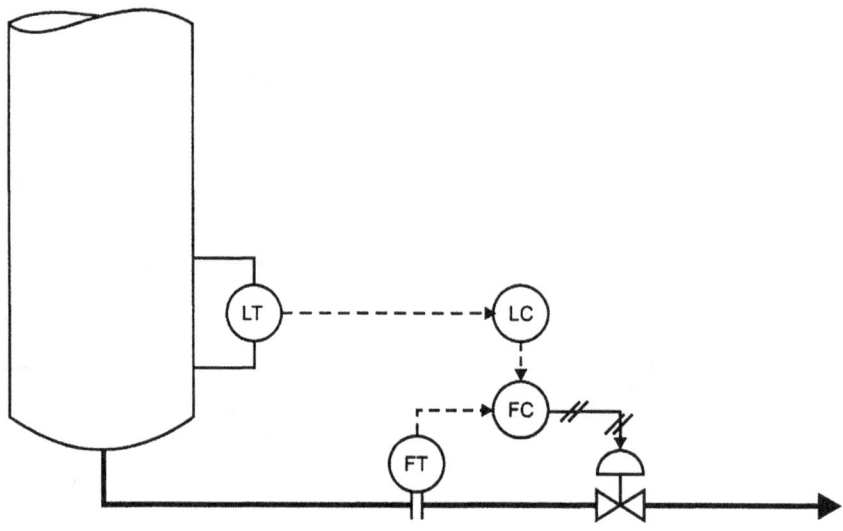

Exercise 10-6.

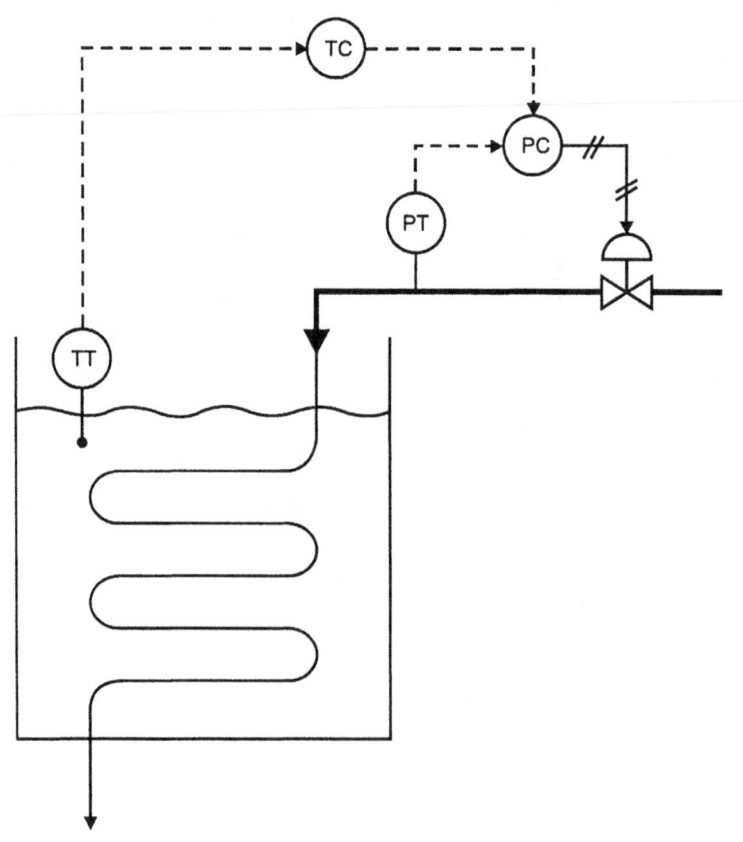

Exercise 10-7.

You are still better off because the inner loop is faster.

Exercise 10-8.

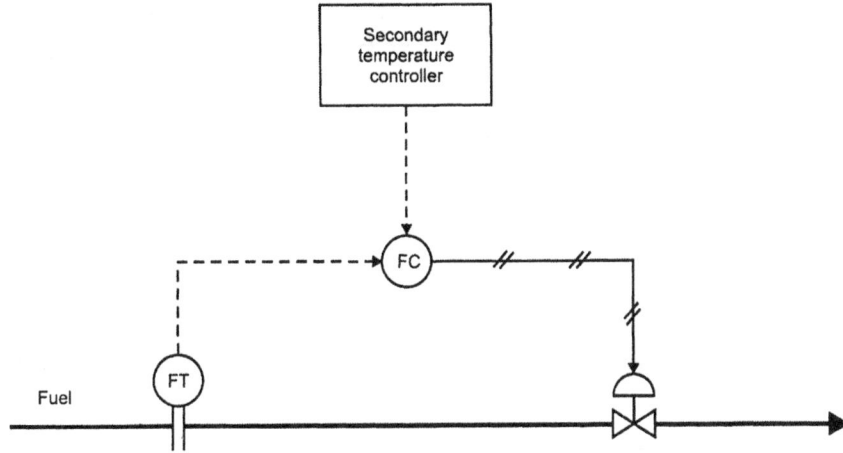

Unit 11

Exercise 11-1.

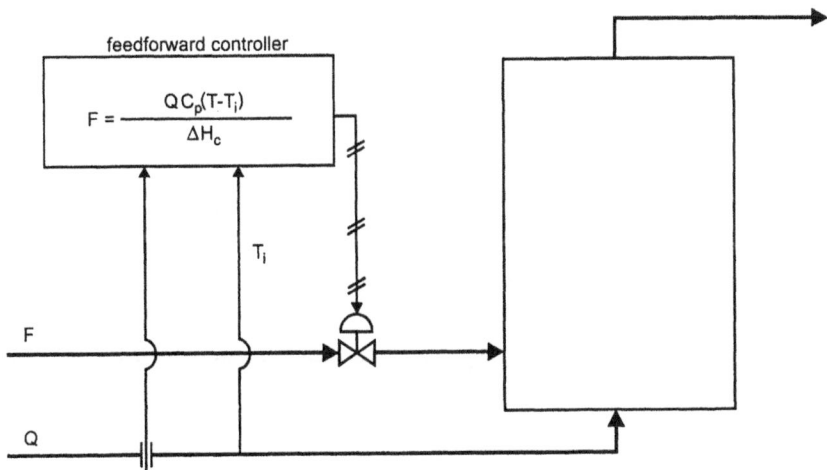

Exercise 11-2.

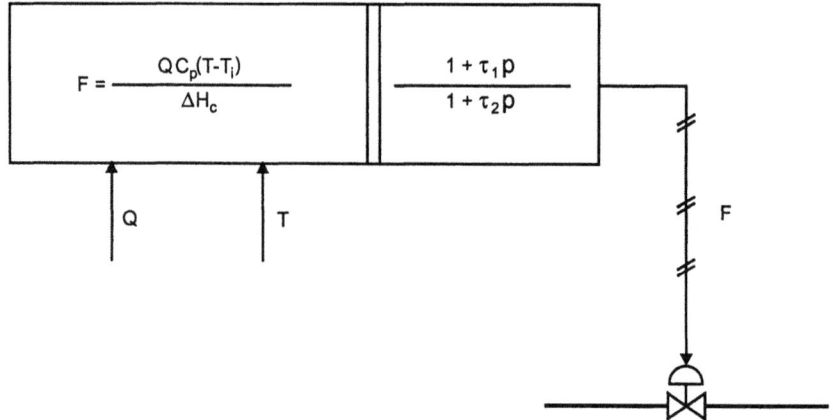

Tuning adjustment can vary τ_1/τ_2 to adjust the dynamic compensation of F.

Exercise 11-3.

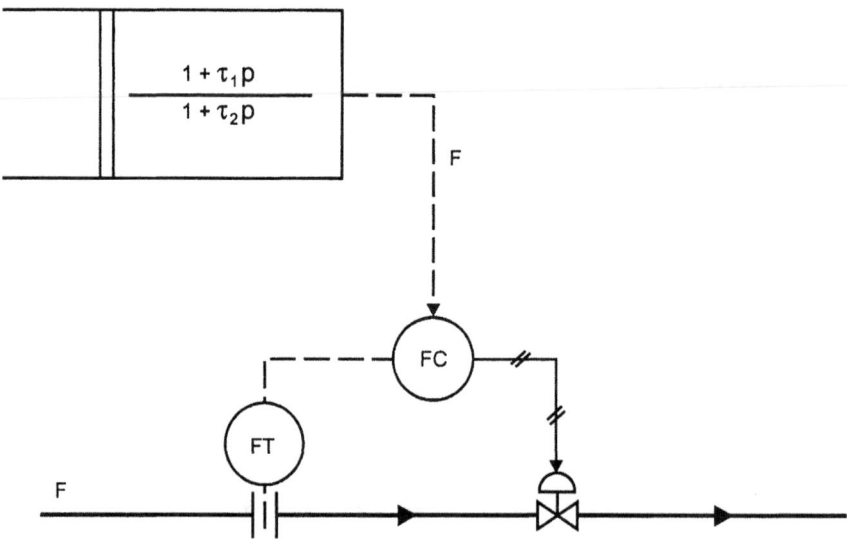

Exercise 11-4.

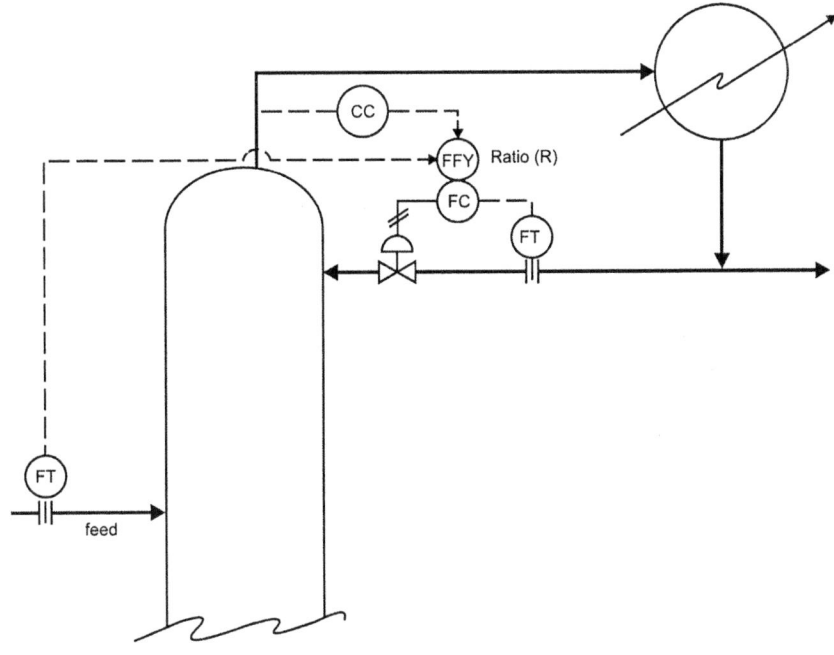

Exercise 11-5.

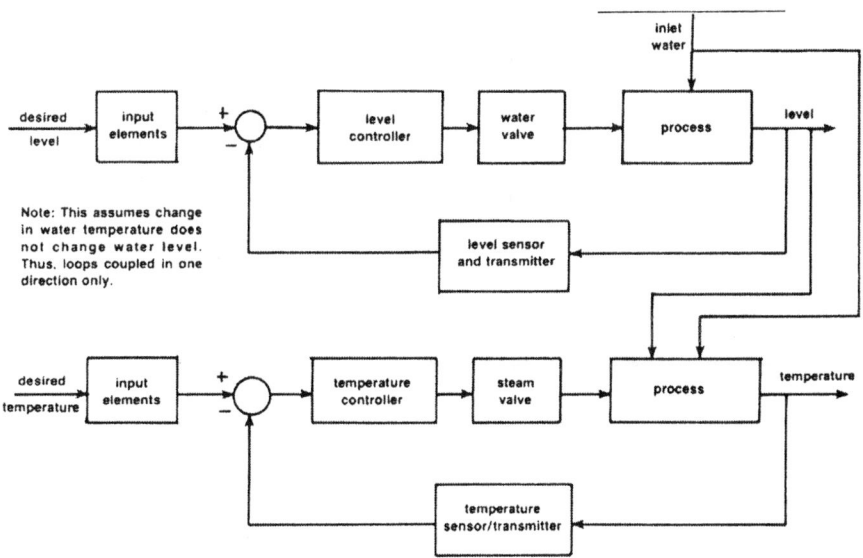

Exercise 11-6.

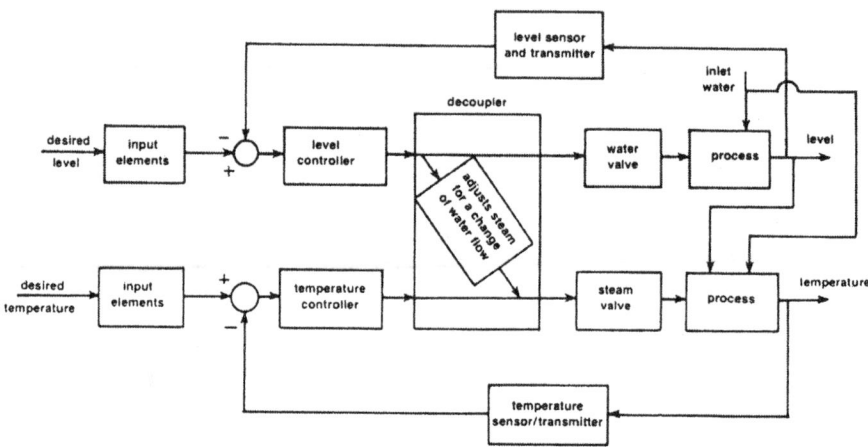

Unit 12

Exercise 12-1.

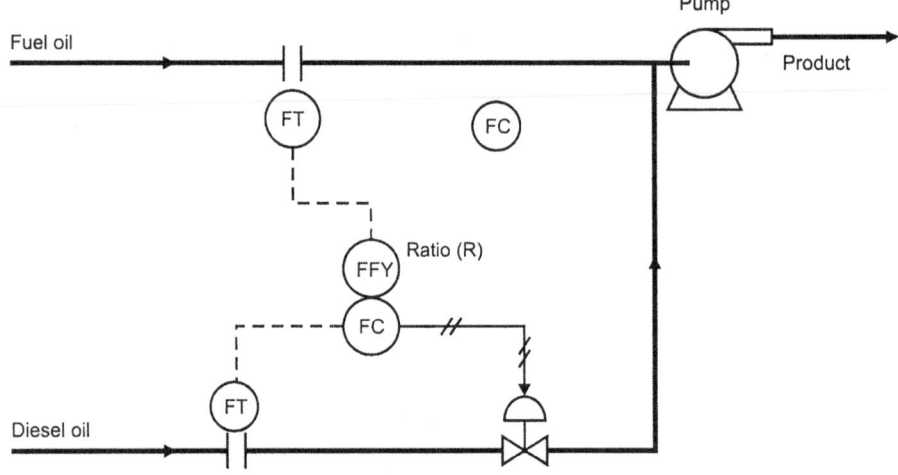

Exercise 12-2.

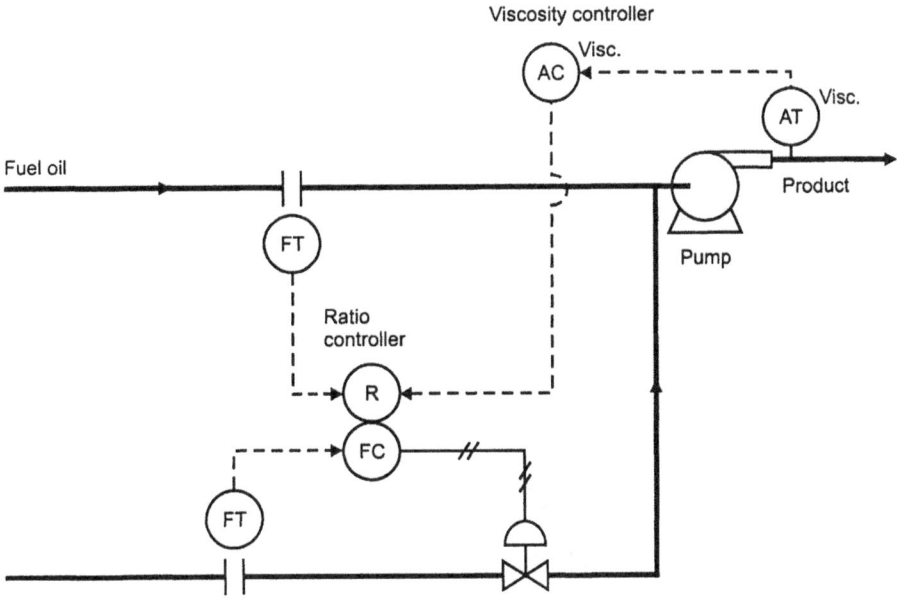

Exercise 12-3.

Liquid level is affected by total flow. Composition is not affected by the absolute value of either flow, only by their ratio.

Exercise 12-4.

To minimize the effect of the composition controller on liquid level, flow B should be the smaller of the two.

Exercise 12-5.

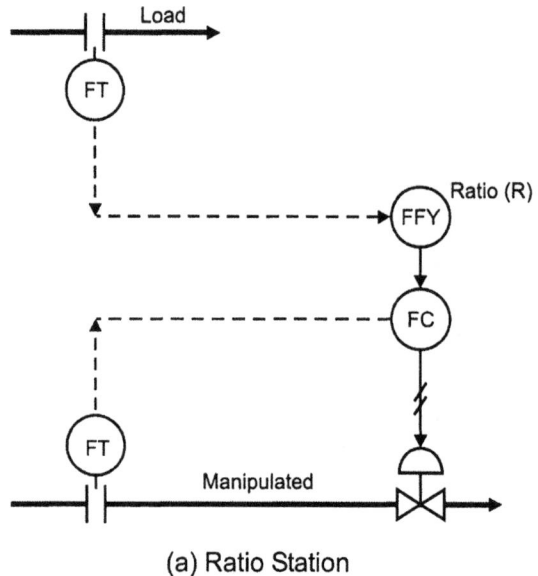

(a) Ratio Station

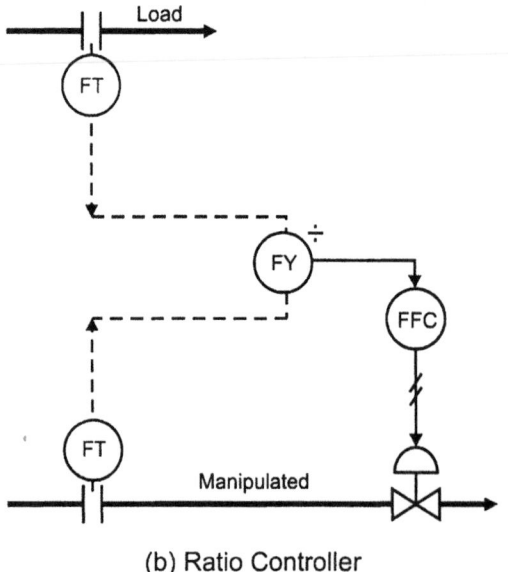

(b) Ratio Controller

Exercise 12-6.

The high selector always selects the transmitter with the highest output, and, in so doing, the controller process variable is always the highest temperature.

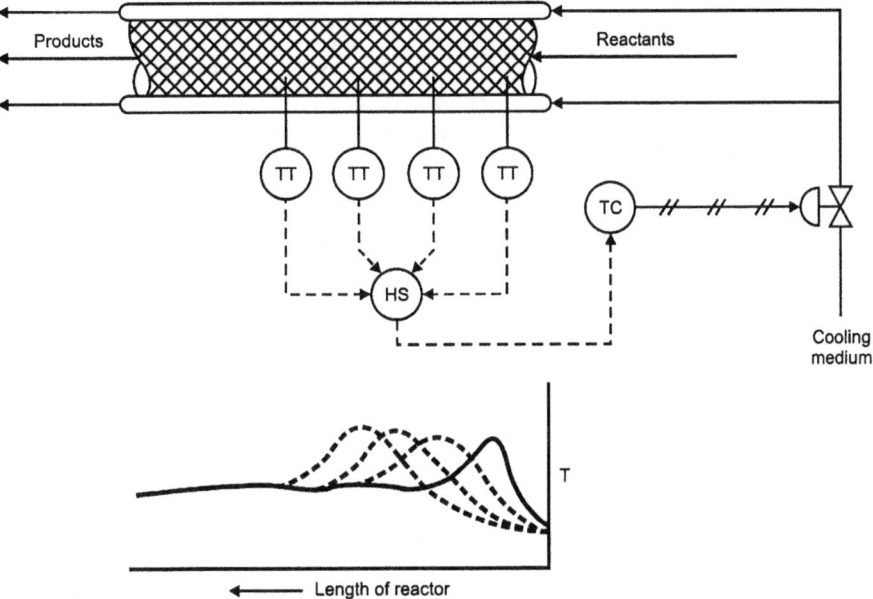

For more detail, see Smith, C. A., and Corripio, A. B., *Principles and Practices of Automatic Control*, John Wiley & Sons, 1985.

Exercise 12-7.

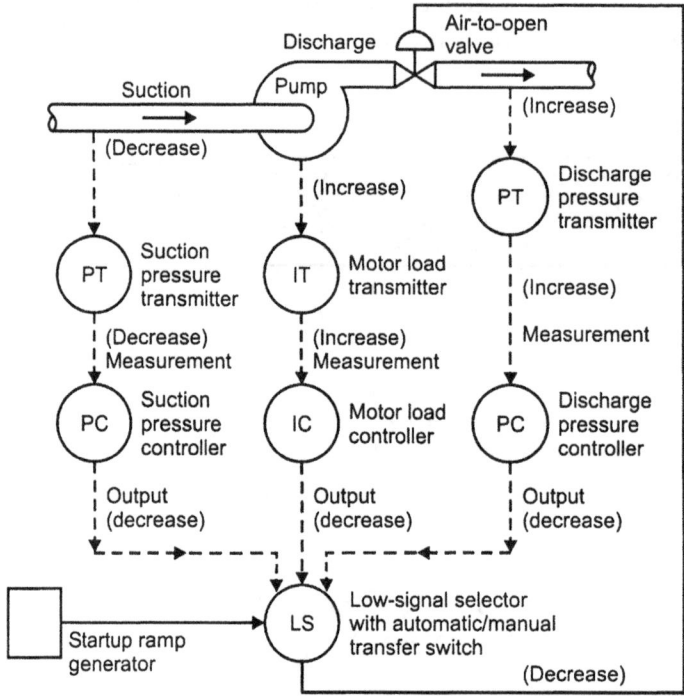

For more detail, see Anderson, N. A., *Instrumentation for Process Measurement and Control*, Third Edition, Chilton Company, 1980.

Unit 13

Exercise 13-1.

The viscosimeter should take its sample at the pump outlet; this lets the impeller do the mixing.

Exercise 13-2.

In a real sense, it is a transportation lag of distance traveled divided by the velocity of the sheet. It is a pure dead time and represents a very difficult control problem of major economic importance. Several vendors have special software packages for each of these installations.

Exercise 13-3.

Most literature comparisons of feedback control with dead time compensators assume PI feedback control. If the modeling is very good, then the predictors tend to be excellent; if the modeling is poor, the predictors do very poorly. The predictors tend to do better for set point changes. The predictors do better for pure dead times (again, assuming good modeling) than for dead time approximations for higher-order processes. For large dead time to time constant ratios, the predictors do better.

Exercise 13-4.

Higher-order approximations can be used, but the mathematics gets more complicated. The extra trouble is only merited if modeling accuracy demands a higher-order approximation.

Exercise 13-5.

A simple technique is to enter a step input to the open-loop process:

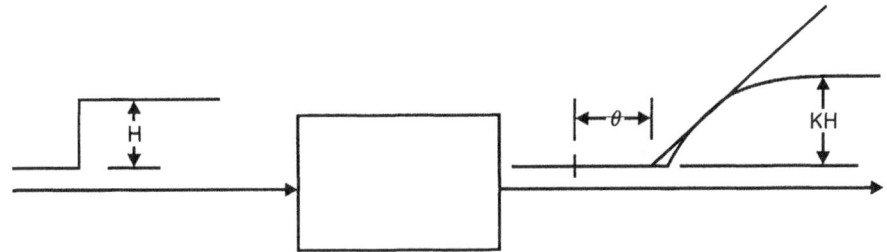

K can be determined by the height of output divided by the height of input. Draw a tangent to output at the point of steepest slope: where it intersects base line is θ. Time constant τ occurs when output reaches 63 percent of its final value.

Unit 14

Exercise 14-1.

Differentiating the controlled variable T_0 with respect to the manipulated variable W gives the following:

$$\frac{dT_o}{dF} = \frac{\Delta H}{WC_p}$$

The gain varies inversely with liquid flow.

Exercise 14-2.

Given the flow characteristic curve for the valve, a compensator is needed whose characteristics are the reverse of those of the valve. For example, a compensator was available in the chosen controller that approximated the curve by five straight-line segments. The segments need not be of equal length, so most of them can be in that portion of the curve where you expect to operate.

Exercise 14-3.

It can be rewritten to reduce the adaptive terms to two:

$$m = \frac{100}{PB}\left(w\,e + \frac{w^2}{T_i} \int e\,dt + T_d \frac{de}{dt} \right)$$

It can be implemented in most modern electronic controllers.

Exercise 14-4.

The answer is vendor dependent.

Other Books by Paul W. Murrill

ISA:

Application Concepts of Process Control. Research Triangle Park, NC: ISA, 1988.

Fundamentals of Process Control Theory. Research Triangle Park, NC: 1981, 2d ed., 1991.

Other Publishers:

Automatic Control of Process. Scranton, PA: International Textbook Company, 1967.

The Formulation and Optimization of Mathematical Models, with R. W. Pike and Cecil L. Smith. Scranton, PA: International Textbook Company, 1970.

BASIC Programming, with Cecil L. Smith. New York: Intext Educational Publishers, 1971. Also published in Spanish as *Programacion Basic*, Ed. Representaciones Y Servicios de Ingenieria, S. A., Intext Educational Publishers, Mexico, 1972.

COBOL Programming, with Cecil L. Smith. New York: Intext Educational Publishers, 1971, 2d ed., 1974.

FORTRAN IV Programming for Engineers and Scientists, with Cecil L. Smith. Scranton, PA: International Textbook Company, 1973.

An Introduction to Computer Science, with Cecil L. Smith. New York: Intext Educational Publishers, 1973. Also published by Intertext Books, International Textbook Company Limited, Aylesbury, Bucks, England, 1973.

An Introduction to FORTRAN IV Programming—A General Approach, with Cecil L. Smith. Scranton, PA: International Textbook Company, 1970. Second edition, New York: Intext Educational Publishers and Thomas Y. Crowell Company, 1975. By 1976 Crowell had acquired Intext.

PL/I Programming, with Cecil L. Smith. New York: Intext Educational Publishers, 1973.

Index